世界最伟大的鸟类图谱珍藏

The Bible of Birds
鸟类圣经

JOHN JAMES AUDUBON 1785-1851

[美] 约翰·詹姆斯·奥杜邦 / 绘

艺术大师编辑部 / 译

北京联合出版公司
Beijing United Publishing Co.,Ltd.

On Stone by Max Rosenthal.

Wild Turkey.

Male.

Drawn from Nature by J.J. Audubon, F.R.S. F.L.S.

Lith. & col. Bowen & Co. Philada.

Wilson's Flycatching-Warbler
Snake's Head Chelone Glabra.

图书在版编目（CIP）数据

乌类圣经/（美)奥杜邦绘；艺术大师编辑部译.
--北京：北京联合出版公司，2015.6
（最伟大的图谱）
ISBN 978-7-5502-4529-7

Ⅰ. ①乌… Ⅱ. ①奥… ②艺… Ⅲ. ①乌类－图集
Ⅳ. ①Q959.7-64

中国版本图书馆CIP数据核字(2015)第132740号

鸟类圣经

项目策划 紫图图书ZITO®
丛书主编 黄利 监制 万夏

作 者 [美] 约翰·詹姆斯·奥杜邦
责任编辑 李 伟
装帧设计 紫图图书ZITO®
封面设计 紫图装帧

北京联合出版公司出版
（北京市西城区德外大街83号楼9层 100088）
北京瑞禾彩色印刷有限公司 新华书店经销
55千字 889毫米×1194毫米 1/16 19.5印张
2015年6月第1版 2015年6月第1次印刷
ISBN 978-7-5502-4529-7
定价：199.00元

目录

如何使用本书

本书以奥杜邦精湛的绘画技术再现了美洲大陆的奇妙物种。奥杜邦的鸟类图谱是图谱类绘画的经典样板，它不但科学、系统，单就每幅画来说，也是值得仔细欣赏的艺术佳作。通过对下面的一个典型页面的分析，使读者能更加轻松地阅读本书。

本书的色彩样板

本书在介绍鸟类时，其色彩的描述依据了自然状态中的颜色和图谱绘制者的原著。其印刷后可能出现的色彩偏差可参见右图所示色彩样板进行校订。

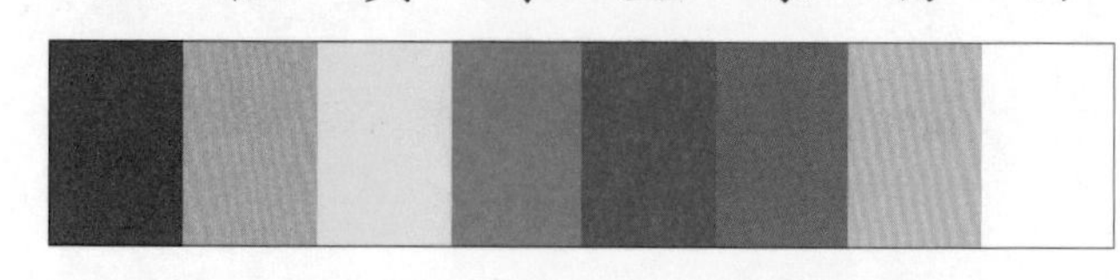

奥杜邦的画像

奥杜邦的鸟类图谱在2002年被选为美国国宝系列邮票的主题之一，它作为美国国宝的地位再次得以确立。

美国国宝级鸟类画家

在美国，约翰·詹姆斯·奥杜邦（1785—1851）可谓家喻户晓，他的《美洲鸟类》一书被认为是19世纪最伟大的著作之一，而且一版再版，至今仍然频繁地被翻印。毫不夸张地说，19世纪的美国文化领域都深深地留下了奥杜邦的烙印。如今的美国，奥杜邦这个名字已经演变为一个包含野生动物、生态保护等含义的文化代名词。他所留下来的画作早已被美国人视为国宝。

宾州费城郊外奥杜邦曾经住过的庄园

约翰·詹姆斯·奥杜邦1785年生于海地，是一位法国船长和他的法国情妇的私生子。幼年的奥杜邦在继母的照顾下生活于法国。为了躲避拿破仑皇帝军队的征召，18岁的奥杜邦远渡重洋来到了美国。父亲在美国的庄园位于费城，在那里他遇到了将来的妻子露西。庄园的小树林让他着迷，树林中数不胜数的鸟类及各种野生动物让他目不暇接，尤其是美洲的鸟类，更是让他痴迷。奥杜邦整天徜徉在青山绿水之间，用他所有的时间来赏鸟和画鸟。在那个汽车尚未发明的年代，他有时骑马，有时坐船，更多时候只能靠着自己的双脚，遍访名山大川，来寻找鸟类的踪迹。那时照相机和望远镜尚未发明，他只能凭着自己的观察，练就高超的绘画技巧。

奥杜邦的妻子露西说："每一只鸟都是我的情敌。"的确如此。奥杜邦对鸟的狂热已经超出了常人

奥杜邦当年所住的房子已经被现代化的高楼大厦所代替，它现在是一个奥杜邦纪念馆。

能够接受的范围，难怪他的妻子心生怨言。“我一直在工作，我真希望自己有八只手来画鸟。”对鸟的痴迷使他无心他顾，以致于连生计都难以维持。而全家的花费，全靠露西做家庭教师的工资来维持。在奥杜邦34岁那年，他甚至因为还不起欠债而被关进监狱，等法院判定他破产，他才被释放出狱。那时奥杜邦仅剩下身上的一套破衣服，一把猎枪，以及一大叠的鸟画。晚年的达尔文回忆起奥杜邦时这么说：“衣服粗糙简单，黝黑的头发在衣领边披散开来，整个人就是一个活脱脱的鸟类标本。”

1826年，奥杜邦带着部分作品来到了英国，

Mrs Audubon
St Francisville
Bayou Sarah
Louisiana

奥杜邦于1827年5月28日写给夫人的信函，上面有他的亲笔签名。

奥杜邦绘制的野火鸡至今为人们所喜爱，这只用它装饰的碟子华丽、精美。

世界上最受欢迎的图谱

奥杜邦的鸟类图谱已成为国际艺术品拍卖市场上最热门的收藏品之一。由奥杜邦签名的原作，价值都在百万美元之上，而后期的再版也在十几万美元以上。如一幅“野火鸡”售价达18.5万美元。

这是《美洲鸟类》第一次印刷之后的一个再次印刷版本中的一页，约制作于1856—1871年。

这幅后人根据奥杜邦的作品重新刻板上色的图谱，售价达1.4万美元。

在奥杜邦之前，亚历山大·威尔逊已经享誉整个美国。威尔逊的图谱（左图）虽然形象准确，但缺乏系统性，鸟类的姿态也欠缺生动。右图是约翰·高德于1849—1861年绘制的图谱，和奥杜邦的风格如出一辙。

世界上最昂贵的书

1992年4月，奥杜邦的《美洲鸟类》原版书，在纽约克里斯蒂公开拍卖，拍价达407万美元之巨，并被誉为世界上最昂贵的书籍之一。《美洲鸟类》自从问世之后就不停地再版和翻印，其中到底出版了多少次，已经无法精确地统计。但至少有一点可以肯定，他的《美洲鸟类》至今仍在以不同的版本形式出现，而且仍然深受世界各地的人们的欢迎。

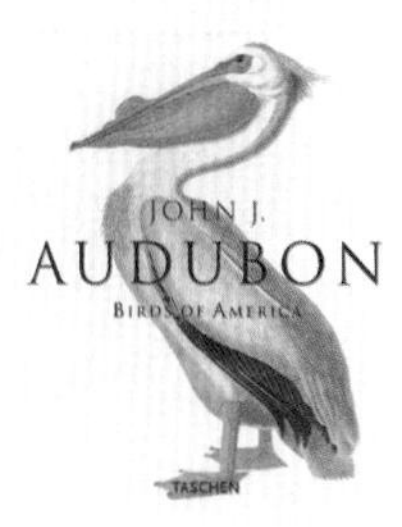

根据奥杜邦的作品所出版的书籍的封面

在伦敦找到了一家印刷厂。随后，一只栩栩如生的野火鸡“飞”了出来，那是奥杜邦的第一幅印刷品。从第一张“野火鸡”的问世，到整本巨著《美洲鸟类》全部完工，整整经历了十二年的时间。此时他已经得到了人们的广泛认可。人们称他为“美国的林中居民”。

奥杜邦的妻子和他的两个女儿

奥杜邦55岁时又开始积极地收集资料，着手绘制美洲的哺乳动物。60岁时，他的视力逐渐恶化，但仍然坚持画画。1848年，《美洲的四足动物》终于绘制完成，他的身体状况也越来越差，三年之后就与世长辞。

1887年的《奥杜邦杂志》封面

奥杜邦的一生就像一场悲喜剧，跌宕起伏。

美国新奥尔良市的奥杜邦动物园

“我希望人们能喜欢它们，我希望人们能够学习尊重大自然，和其他动物们和谐相处。”奥杜邦在他的日记中对人们滥杀鸟类、破坏生态环境的行为发出了警告。到了19世纪末期，人们才开始去思考奥杜邦的理念。1887年，在奥杜邦死后三十多年，人们建立了以他的名字命名的团体——奥杜邦学会。奥杜邦学会现在已经成为全球最具影响力的环保团体之一。它拥有55万会员，508个分支机构，100处馆舍，还有300位包括科学家、教育家在内的专职工作人员为它工作，大批志愿工作者和100多个保护区遍布美国。

奥杜邦博物馆

美国俄亥俄州雅典市的奥杜邦博物馆及自然学家中心，是北美地区数量庞大的以奥杜邦的名字命名的科学馆所之一。该中心以奥杜邦及其家庭的故事为布展线索，集中展示了奥杜邦作品最早的印刷品。这个博物馆有鸟类解剖学及行为学方面的展览，人们在这里还有机会观察和记录鸟类在原生栖息地的状态。这个博物馆在某种程度上复原着奥杜邦倡导的自然人文精神。

奥杜邦的另一杰作

1854年，纽约V.G.奥杜邦出版社出版了奥杜邦的另一部专著《美洲的四足动物》。作者以一贯的生动风格和精细画风，描绘了美洲的150种四足动物，该著作为奥杜邦赢得了更大的盛誉。

这是《美洲的四足动物》中一幅美洲猫图谱。

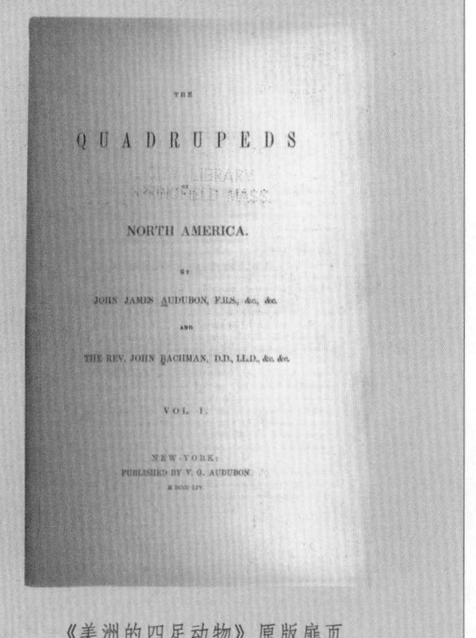

THE
QUADRUPEDS
OF
NORTH AMERICA.
BY
JOHN JAMES AUDUBON, F.R.S., &c. &c.
AND
THE REV. JOHN BACHMAN, D.D., LL.D., &c. &c.
VOL. I.
NEW-YORK:
PUBLISHED BY V. G. AUDUBON.
MDCCCLIV.

《美洲的四足动物》原版扉页

西方最流行的装饰画

奥杜邦的作品已经被美国人视为国宝级的文化遗产。自从奥杜邦的作品问世以来，以鸟类画作为装饰就风靡美国乃至整个西方。以鸟类图谱装饰自己的居室在19世纪和20世纪被视为一种时尚。绘有奥杜邦的图谱图案的瓷器已经成为古董收藏家的珍藏。2002年，美国发行了一套美国国宝系列邮票，其中就有奥杜邦的鸟类图谱。

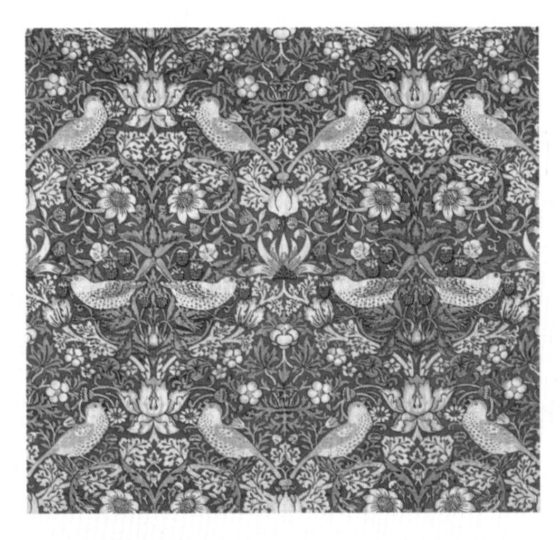

这是引导20世纪装饰艺术潮流的新艺术运动大师威廉·莫里斯设计的布纹图案，他的鸟类造型深受奥杜邦绘画风格的影响。

绘有奥杜邦的鸟类图谱的咖啡杯。

绘有奥杜邦的鸟类图谱的碟子。

这是新艺术运动大师威廉·莫里斯在1890年前后设计和装饰的一处住宅，墙上所挂的两幅鸟类装饰画，正是出自奥杜邦之手。当时用鸟类图谱做装饰已经成为一种风尚。

再现杰作

2002年8月，南印第安那州大学的艺术学教授迈克·阿克尤斯复制了奥杜邦435幅鸟类作品中的一幅。奥杜邦的《美洲鸟类》于1839年在伦敦全部印刷完成，尔后这些铜版被运送到纽约，从此这些铜刻板开始遗失。迈克·阿克尤斯复制作品所用的这块铜质雕版，是奥杜邦的《美洲鸟类》最早的版本所仅存的78块之一，这是这块铜质雕版在160年后，首次拓印作品。这块雕版曾被奥杜邦的朋友收藏，现收藏于肯塔基州的奥杜邦博物馆。到目前为止，全美只有五家机构复制过原作。

在拓印之前，迈克·阿克尤斯教授先轻轻地擦去铜版上的油污。

拓印好的成品

着色后的图谱

迈克·阿克尤斯教授从铜版上揭下拓印的版画。

在南印第安那州大学的艺术工作室，肯塔基州奥杜邦博物馆的馆长鲍曼先生向人们展示了奥杜邦的鸟类图谱的铜版制画步骤。

Californian Turkey Vulture.

Drawn from Nature by J.J.Audubon, F.R.S.F.L.S.

Lith. Printed & Col^{d} by J.T.Bowen, Phil^{a}

加利福尼亚美洲鹫

学名：Cathartes aura　英文名：Californian Turkey Vulture

Red-headed Turkey Vulture.

Drawn from Nature by J.J.Audubon, F.R.S.F.L.S

Lith. Printed & Col.d by J.T.Bowen, Phil

红头美洲鹫

学名：Cathartes aura　英文名：Red-headed Turkey Vulture

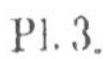

Black Vulture or Carrion Crow.

Drawn from Nature by J.J.Audubon, F.R.S.F.L.S.

黑秃鹰

学名：Coragyps atratus　英文名：Black Vulture or Carrion Crow

Pl. 14

Drawn from Nature by J.J.Audubon,F.R.S.F.L.S.

Drawn on Stone by Wm E. Hitchcock.

White-headed Sea-Eagle or Bald-Eagle

白头海雕

学名：*Haliaeetus leucocephalus*　英文名：White-headed Sea Eagle or Bald Eagle

Caracara Eagle.

Drawn from Nature by J.J.Audubon, F.R.S.F.L.S.

Lith. Printed & Col^d by J.T. Bowen, Phila.

长腿兀鹰

学名：Polyborus plancus 英文名：Caracara Eagle

Harris's Buzzard

Drawn from Nature by J. J. Audubon, F.R.S.F.L.S.　　Lith. Printed & Col^{d} by J. T. Bowen Phil.

栗翅鹰　　学名：Parabuteo unicinctus　英文名：Harris's Buzzard

Common Buzzard

Drawn from Nature by J.J.Audubon, F.R.S.F.L.S.

Lith.^d Printed & Col.^d by J. T. Bowen Philad.^a

鵟

学名：Buteo buteo　英文名：Common Buzzard

Red-tailed Buzzard.

Drawn from Nature by J.J.Audubon, F.R.S.F.L.S. Lith.d Printed & Col.d by J. T. Bowen, Philad.a

红尾鵟 学名：Buteo jamaicensis 英文名：Red-tailed Buzzard

Pl. 8.

Harlan's Buzzard.

Drawn from Nature by J.J. Audubon, F.R.S.F.L.S. Lith.d Printed & Col.d by J. T. Bowen, Philad.a

哈兰鹰

学名：Buteo Harlani 英文名：Harlan's Buzzard

红肩鹰

学名：Buteo lineatus 英文名：Red-shouldered Buzzard

Pl 10

Broad winged Buzzard

Drawn from Nature by J.J.Audubon,F.R.S.F.L.S. Lith.^d Printed & Col.^d by J. T. Bowen Philad.^a

宽翅鹰

学名：Buteo platypterus 英文名：Broad-winged Buzzard

Rough-legged Buzzard.

Drawn from Nature by J.J.Audubon,F.R.S.F.L.S.　　Lith.d Printed & Col.d by J. T. Bowen Philad.a

粗腿鹰　　学名：Buteo lagopus　英文名：Rough-legged Buzzard

Drawn from Nature by J.J.Audubon.F.R.S.F.L.S.　　Lith.d Printed & Col.d by J. T. Bowen, Philad.a

金雕　　学名：Aquila chrysaetos　英文名：Golden Eagle

Common Osprey Fish Hawk.

Drawn from Nature by J.J. Audubon, F.R.S.F.L.S. Lith. Printed & Col^{d} by J.T. Bowen, Phila.

鱼鹰 学名：Pandion haliaetus 英文名：Common Osprey or Fish Hawk

Pl. 20.

Drawn from Nature by J.J. Audubon, F.R.S.F.L.S.

Lith. & Col. Bowen & Co Philad.a

Peregrine Falcon

游隼

学名：Falco peregrinus　英文名：Peregrine Falcon

Pl. 18.

Drawn from Nature by J.J. Audubon, F.R.S.F.L.S.

Lith.d Printed & Col.d by J. T. Bowen, Philad.a

Swallow-tailed Hawk.

燕尾鸢

学名：Elanoides forficatus　英文名：Swallow-tailed Hawk

Pl. 16.

Black-shouldered Elanus.

Drawn from nature by J. W. Audubon. Lith. & col. by Bowen & Co. Philad.ª

黑翅鸢 学名：Elanus caeruleus 英文名：Black-shouldered Elanus

密西西比鸢

学名：Ictinia mississippiensis 英文名：Mississippi Kite

Drawn from Nature by J.J.Audubon,F.R.S.F.L.S.　　Lith[d] Printed & Col[d] by J.T.Bowen,Philad.

Iceland or Gyr Falcon

矛隼

学名：Falco rusticolus　英文名：Iceland or Gyr Falcon

Pigeon Falcon.

Drawn from Nature by J.J.Audubon, F.R.S.F.L.S. Lith.d Printed & Col.d by J. T. Bowen, Philad.a

灰背隼

学名：Falco columbarius 英文名：Merlin

Sparrow Falcon.

Drawn from Nature by J.J.Audubon.F.R.S.F.L.S

Lith.d Printed & Col.d by J. T Bowen, Philad.a

雀鹰

学名：Accipiter nisus 英文名：Sparrow Falcon

Gos Hawk.

Drawn from Nature by J.J.Audubon,F.R.S.F.L.S.　　Lith.d Printed & Col.d by J. T. Bowen, Philad.a

苍鹰　　学名：Accipiter gentilis　英文名：Gos Hawk

Pl. 25.

Sharp-shinned Hawk

Drawn from Nature by J.J.Audubon,F.R.S.F.L.S.　　　　Lith.d Printed & Col.d by J. T. Bowen Philad.a

条纹鹰　　　　学名：Accipiter striatus　英文名：Sharp-shinned Hawk

Pl. 26.

Common Harrier.

Drawn from Nature by J.J.Audubon,F.R.S.F.L.S.　　Lith.^d Printed & Col.^d by J. T. Bowen, Philad.^a

灰泽鵟　　学名：Circus cyaneus　英文名：Common Harrier

Hawk Owl.

Drawn from Nature by J.J.Audubon, F.R.S.F.L.S. — Lith.d Printed & Col.d by J. T. Bowen, Philad.a

鹰鸮 学名：Ninox scutulata 英文名：Hawk Owl

Snowy Owl.

Drawn from Nature by J.J.Audubon, F.R.S.F.L.S.　　Lith.d Printed & Col.d by J. T. Bowen, Philad.

雪鸮　　学名：Nyctea scandiaca.　英文名：Snowy Owl

Pl.33.

Little or Acadian Owl

Common Mouse

Drawn from Nature by J.J.Audubon,F.R.S.F.L.S.　　Lith[d] Printed & Col[d] by J. T. Bowen, Philad[a]

拉锯枭　　学名：Aegolius acadicus　英文名：Saw Whet Owl

Barn Owl

Drawn from Nature by J.J. Audubon, F.R.S.F.L.S

Lith. & Col. Bowen & Co. Philad.ª

仓鸮

学名：Tyto alba　英文名：Barn Owl

Great Cinereous Owl.

Drawn from Nature by J.J. Audubon. F.R.S.F.L.S. Lith^d Printed & Col^d by J. T. Bowen, Philada.

灰鸮 学名：Strix nebulosa 英文名：Great Cinereous Owl

Barred Owl

Drawn from Nature by J.J.Audubon,F.R.S.F.L.S.　　Lith.d Printed & Col.d by J.T.Bowen, Philad.a

横斑林鸮

学名：Strix varia　英文名：Barred Owl

Pl.31.

Burrowing Day-Owl.

Drawn from Nature by J.J. Audubon F.R.S.F.L.S. Lith.Printed & Col.d by J.T.Bowen, Philad.a

穴居鸮

学名：Speotyto cunicularia 英文名：Burrowing Owl

Pl.38.

Short-eared Owl.

Drawn from Nature by J.J.Audubon F.R.S.F.L.S. Lith.Printed & Col.d by J.T.Bowen, Philad.a

短耳鸮

学名：Asio flammeus 英文名：Short-eared Owl

Great Horned-Owl.

Drawn from Nature by J.J.Audubon, F.R.S.F.L.S.

Lith.d Printed & Col.d by J. T. Bowen, Philad.a

巨角鸮

学名：Bubo virginianus　英文名：Great Horned-Owl

Little Screech Owl.

Jersey Pine. Pinus inops.

Drawn from Nature by J.J. Audubon, F.R.S.F.L.S.

Lith.d Printed & Col.d by J. T. Bowen, Philad.a

小长耳鸮

学名：Strix asio　英文名：Little Screech-Owl

Pl. 43.

Night Hawk

White Oak Quercus Alba

Drawn from Nature by J.J.Audubon, F.R.S.F.L.S.　　　Lith.d Printed & Col.d by J. T. Bowen, Philad.a

夜鹰　　　学名：Caprimulgus indicus　英文名：Night-Hawk

Purple Martin.

(Calabash.)

Drawn from Nature by J.J.Audubon,F.R.S.F.L.S

Lith^d Printed & Col^d by J. T. Bowen, Philad^a

北美洲紫燕

学名：Progne subis 英文名：Purple Martin

White-bellied Swallow.

Drawn from Nature by J.J.Audubon,F.R.S.F.L.S.　　Lith.d Printed & Col.d by J. T. Bowen Philad.a

树燕　　学名：Tachycineta bicolor　英文名：White-bellied Swallow

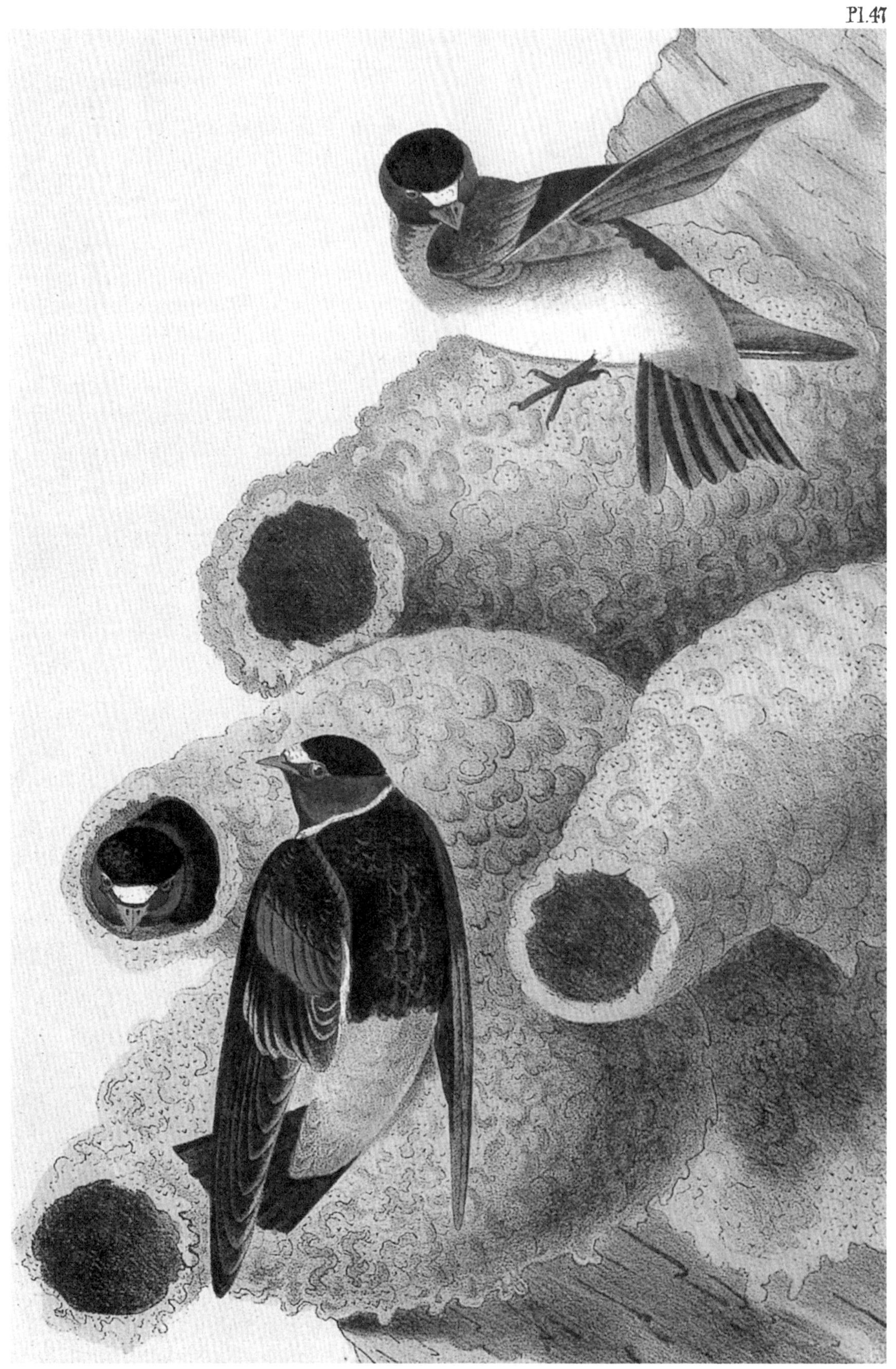

Cliff Swallow.

(Nests.)

Drawn from Nature by J.J.Audubon.F.R.S.F.L.S

Lith.d Printed & Col.d by J.T.Bowen Philad.a

岩燕

学名：Hirundo pyrrhonota 英文名：Cliff Swallow

Barn or Chimney Swallow

Drawn from Nature by J.J.Audubon.F.R.S.F.L.S.　　Lith.d Printed & Col.d by J. T Bowen. Philad.a

家燕　　学名：Hirundo rustica　英文名：Barn or Chimney Swallow

Violet Green Swallow

Drawn from Nature by J.J.Audubon, F.R.S.F.L.S.

Lith, Printed & Col^d by J.T. Bowen, Phil^a

紫绿燕

学名：Tachycineta thalassina　英文名：Violet-green Swallow

Male.

Drawn from Nature by J.J.Audubon.F.R.S.F.L.S.　　Lith.d Printed & Col.d by J. T. Bowen, Philad.a

黄腹捕蝇鸟（雄性）　　学名：Empidonax flaviventris　英文名：Yellow-bellied Flycatcher

Fork-tailed Flycatcher.

Gordonia Lasianthus.

Drawn from Nature by J.J.Audubon, F.R.S.F.L.S. Lith.d Printed & Col.d by J. T. Bowen Philad.a

叉尾捕蝇鸟

学名：Tyrannus savana 英文名：Forked-tailed Flycatcher

Pl 53

Swallow-tailed Flycatcher.

Drawn from Nature by J.J.Audubon,F.R.S.F.L.S. Lith[d] Printed & Col[d] by J. T. Bowen Philad[a]

剪尾王霸鹟

学名：Tyrannus forficatus 英文名：Swallow-tailed Flycatcher

Arkansas Flycatcher.

Drawn from Nature by J.J.Audubon, F.R.S.F.L.S.

Lith.d Printed & Col.d by J. T. Bowen, Philad.a

阿肯色捕蝇鸟

学名：Tyrannus flaviventris 英文名：Arkansas Flycatcher

Tyrant Flycatcher or King Bird.

Cotton wood. Populus candicans.

Drawn from Nature by J.J.Audubon,F.R.S.F.L.S.

Lith.d Printed & Col.d by J. T. Bowen, Philad.a

王鸟

学名：Tyrannus tyrannus　英文名：Tyrant Flycatcher or King-Bird

Pl. 57.

Great Crested Flycatcher

Drawn from Nature by J.J.Audubon.F.R.S.F.L.S.　　　Lith. Printed & Col.d by J. T. Bowen, Philad.a

凤头捕蝇鸟　　　学名：Myiarchus crinitus　英文名：Great Crested Flycatcher

Say's Flycatcher

1. Male 2. Female.

Drawn from Nature by J.J.Audubon,F.R.S.F.L.S.

Lith^d Printed & Col^d by J. T. Bowen, Philad^a

菲比霸鹟（1雄性，2雌性）

学名：Sayornis saya 英文名：Say's Flycatcher

Rocky Mountain Flycatcher.
(Swamp Oak Quercus Aquatica.)
Male.

Drawn from Nature by J.J.Audubon,F.R.S.F.L.S.

Lith.d Printed & Col.d by J. T. Bowen, Philad.a

山岩捕蝇鸟（雄性）

学名：Sayornis nigricans　英文名：Rocky Mountain Flycatcher

Small Green-crested Flycatcher.

Sassafras Laurus Sassafras.

1. Male 2. Femal.

Drawn from Nature by J.J. Audubon, F.R.S.F.L.S.

Lith.d Printed & Col.d by J. T. Bowen, Philad.a

阿卡迪亚捕蝇鸟（1雄性、2雌性）

学名：Empidonax virescens　英文名：Small Green-crested Flycatcher

Pewee Flycatcher.

Cotton Plant. Gossypium Herbaceum.

1. Male 2. Female.

Drawn from Nature by J.J.Audubon, F.R.S.F.L.S. | Lith.d Printed & Col.d by J. T. Bowen, Philad.a

美洲小燕雀（1雄性、2雌性）

学名：Empidonax minimus　英文名：Pewee Flycatcher

Pl. 65.

Drawn from Nature by J.J.Audubon,F.R.S.F.L.S

Lith.d Printed & Col.d by J. T. Bowen, Philad.a

Traill's Flycatcher.
Male.
Sweet Gum. Liquidambar Styracifolia.

跟踪捕蝇鸟（雄性）

学名：Empidonax traillii　英文名：Traill's Flycatcher

Least Pewee Flycatcher.

White Oak. Quercus Prinus.

Male.

Drawn from Nature by J.J.Audubon, F.R.S.F.L.S.

Lith.d Printed & Col.d by J. T. Bowen, Philad.a

小燕捕蝇鸟（雄性）

学名：Empidonax minimus 英文名：Least Pewee Flycatcher

Pl. 67.

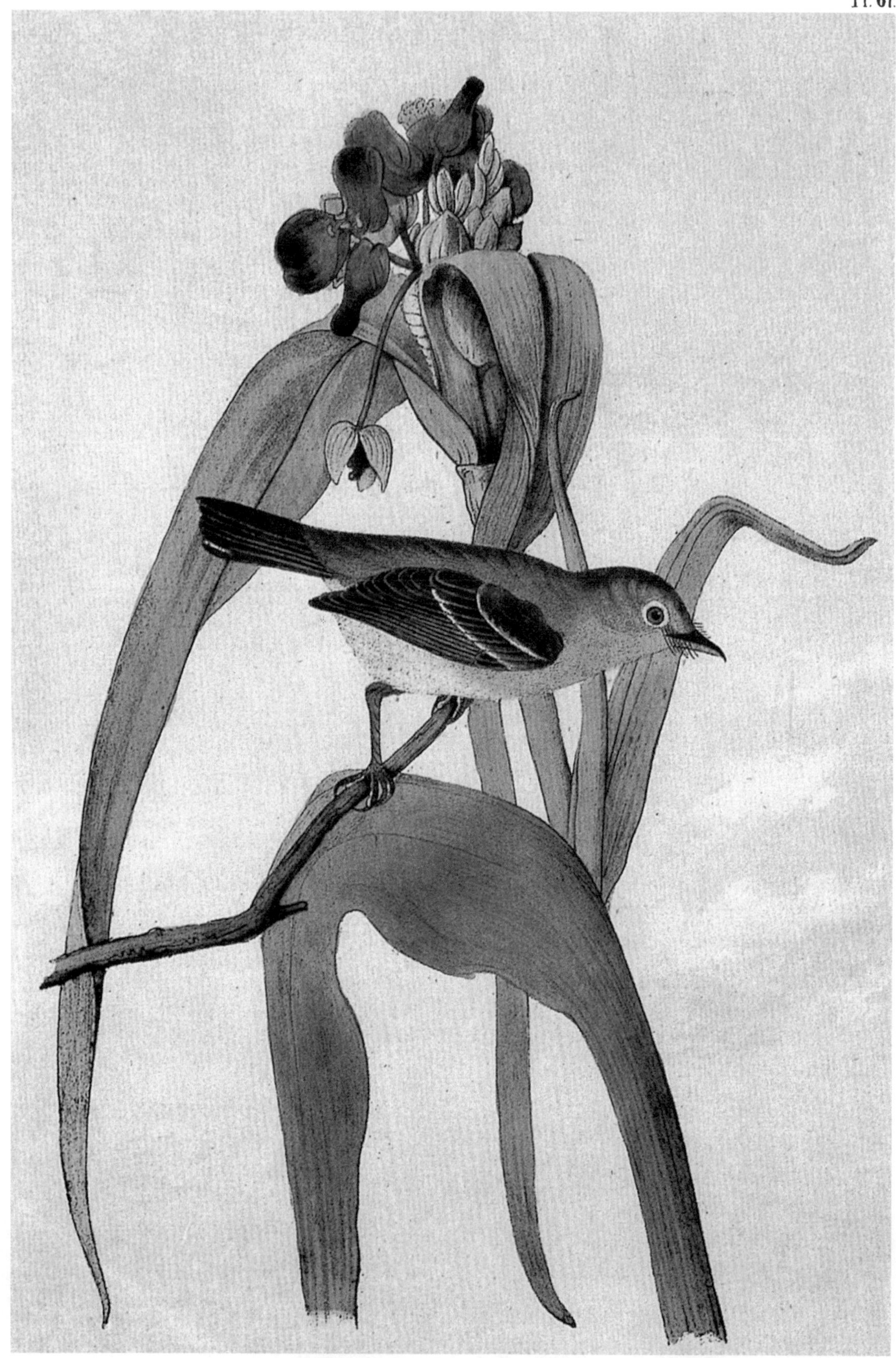

Small-headed Flycatcher

Virginian Spider-wort. Tradescantia Virginica.

Male.

Drawn from Nature by J.J.Audubon.F.R.S.F.L.S. Lith.d Printed & Col.d by J. T. Bowen, Philad.a

小头捕蝇鸟（雄性）

学名：Sylvania microcephala 英文名：Small-headed Flycatcher

Pl. 68.

American Redstart.

Virginian Hornbeam or Iron-wood Tree.

1. Male 2. Female.

Drawn from Nature by J.J.Audubon,F.R.S.F.L.S. Lith^d Printed & Col^d by J. T. Bowen, Philad^a

美洲红尾鸟（1雄性、2雌性） 学名：Tyrannus verticallis 英文名：American Redstart

Pl. 69.

Townsend's Ptilogonys.

Female.

Drawn from Nature by J.J.Audubon,F.R.S.F.L.S. Lith.d Printed & Col.d by J. T. Bowen, Philad.a

汤森宝石（雄性）

学名：Myadestes townsendi 英文名：Townsend's Ptilogonys

Blue-grey Flycatcher

Black Walnut Juglans nigra.

1 Male 2 Female.

Drawn from Nature by J.J.Audubon,F.R.S.F.L.S.

Lith.d Printed & Col.d by J. T. Bowen, Philad.a

灰蓝色捕蝇鸟（1雄性、2雌性）

学名：Polioptila caerulea 英文名：Blue-grey Flycatcher

Pl. 71.

Hooded Flycatching Warbler

Erithryna herbacea.

1.Male 2.Female.

Drawn from Nature by J.J.Audubon,F.R.S.F.L.S.

Lith.d Printed & Col.d by J. T. Bowen, Philad.a

头巾林莺（1雄性、2雌性）

学名：Wilsonia citrina　英文名：Hooded Flycatching-Warbler

Canada Flycatcher

Great Laurel Rhododendron maximum

1 Male 2 Female

Drawn from Nature by J.J.Audubon,F.R.S.F.L.S.

Lith.d Printed & Col.d by J. T. Bowen, Philad.a

加拿大捕蝇鸟（1雄性、2雌性）

学名：Wilsonia canadensis 英文名：Canada Flycatcher

Bonaparte's Flycatching-Warbler.

Great Magnolia. Magnolia Grandiflora.

Male.

Drawn from Nature by J.J.Audubon,F.R.S.F.L.S. Lith.d Printed & Col.d by J. T. Bowen, Philad.a

波拿巴林莺（雄性）

学名：Wilsonia canadensis 英文名：Bonaparte's Flycatching-Warbler

Kentucky Flycatching Warbler

Magnolia auriculata.

1. Male. 2. Female.

Drawn from Nature by J.J.Audubon, F.R.S.F.L.S.   Lith.d Printed & Col.d by J. T. Bowen, Philad.a

肯塔基林莺（1雄性、2雌性） 学名：Oporornis formosus 英文名：Kentucky Flycatching-Warbler

Wilson's Flycatching-Warbler.

Snake's Head Chelone Glabra.

1. Male 2. Female.

Drawn from Nature by J.J.Audubon,F.R.S.F.L.S.

Lith.d Printed & Col.d by J. T. Bowen, Philad.a

威尔逊林莺（1雄性、2雌性）

学名：Wilsonia pusilla 英文名：Wilson's Flycatching-Warbler

Pl. 76.

Yellow-crowned Wood-Warbler.

Iris versicolor.

1. Male. 2. Young.

Drawn from Nature by J.J.Audubon.F.R.S.F.L.S.   Lith.d Printed & Col.d by J. T. Bowen, Philad.a

皇冠林莺（1雄性、2幼鸟）

学名：Dendroica coronata　英文名：Yellow-crowned Wood-Warbler

Audubon's Wood-Warbler.

Strawberry Tree Euonymus Americanus.

1. Male. 2. Female.

Drawn from Nature by J.J. Audubon, F.R.S.F.L.S.

Lith.d Printed & Col.d by J. T. Bowen, Philad.a

奥杜邦林莺（1雄性、2雌性）

学名：Dendroica coronata　英文名：Audubon's Wood-Warbler

Black-poll Wood Warbler.

Black Gum Tree. Nyssa aquatica

1. Males 2. Female.

Drawn from Nature by J.J.Audubon, F.R.S.F.L.S.

Lith.d Printed & Col.d by J. T. Bowen, Philad.a

黑顶白颊林莺（1雄性、2雌性）

学名：Dendroica striata　英文名：Black-poll Wood-Warbler

Bay-Breasted Wood-Warbler.

Highland Cotton-plant. Gossipium herbaceum.

1. Male 2. Female

Drawn from Nature by J.J.Audubon,F.R.S.F.L.S. Lith.d Printed & Col.d by J. T. Bowen, Philad.a

海湾胸林莺（1雄性、2雌性） 学名：Dendroica castanea 英文名：Bay-Breasted Wood-Warbler

Pl. 82.

Pine creeping Wood-Warbler

Yellow Pine Pinus variabilis

1. Male 2. Female

Drawn from Nature by J.J. Audubon F.R.S.F.L.S. | Lith. Printed & Col. by J. T. Bowen, Philad.

爬松林莺（1雄性、2雌性） 学名：Dendroica pinus 英文名：Pine Creeping Wood-Warbler

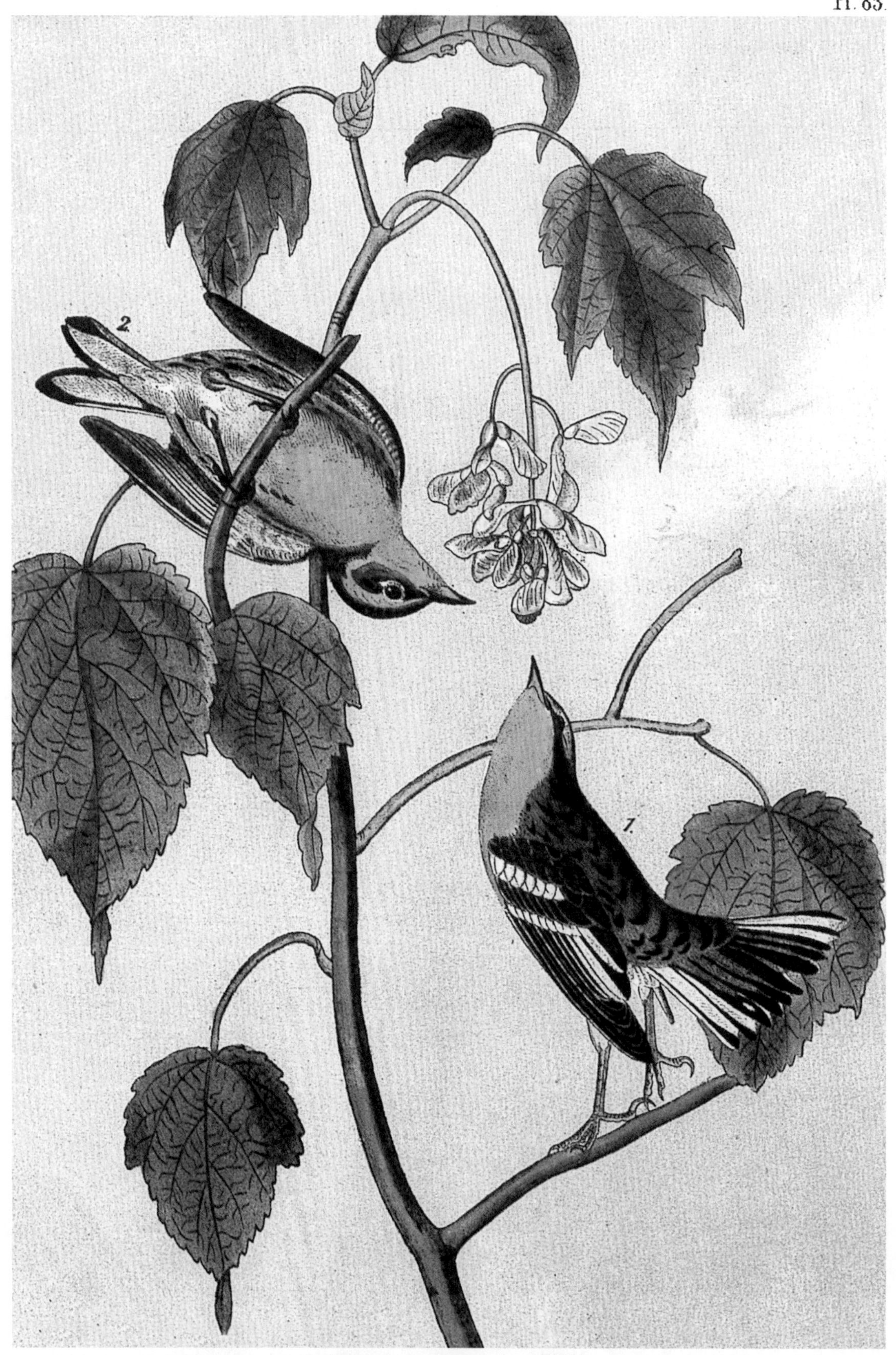

Hemlock Warbler.

Dwarf Maple. Acer Spicatum.

1. Male 2. Female.

Drawn from Nature by J.J. Audubon, F.R.S.F.L.S.

Lith.d Printed & Col.d by J. T. Bowen. Philad.a

芹叶钩吻林莺（1雄性、2雌性）

学名：Dendroica fusca　英文名：Hemlock Warbler

Pl. 84.

Black-throated Green Wood Warbler.

1, Male. 2, Female.

Caprifolium Sempervirens.

Drawn from Nature by J.J.Audubon, F.R.S.F.L.S. | Lith.d Printed & Col.d by J. T. Bowen, Philad.a

黑喉绿林莺（1雄性、2雌性）

学名：Dendroica virens　英文名：Black-throated Green Wood-Warbler

Cape May Wood Warbler.

1. Male 2. Female.

Drawn from Nature by J.J.Audubon, F.R.S.F.L.S.

Lith.d Printed & Col.d by J. T. Bowen, Philad.a

五月林莺（1雄性、2雌性）

学名：Dendroica tigrina　英文名：Cape May Wood-Warbler

Blackburnian Wood Warbler.

1. Male 2 Female.

Phlox maculata.

Drawn from Nature by J.J.Audubon.F.R.S.F.L.S.　　Lith.^d Printed & Col.^d by J. T. Bowen, Philad.^a

黑斑林莺（1雄性、2雌性）　　学名：Dendroica fusca　英文名：Blackburnian Wood-warbler

Yellow-poll Wood Warbler

Males.

Drawn from Nature by J.J.Audubon,F.R.S.F.L.S.   Lith.d Printed & Col.d by J. T. Bowen, Philad.a

黄林莺（雄性）

学名：Dendroica petechia　英文名：Yellow-poll Wood-Warbler

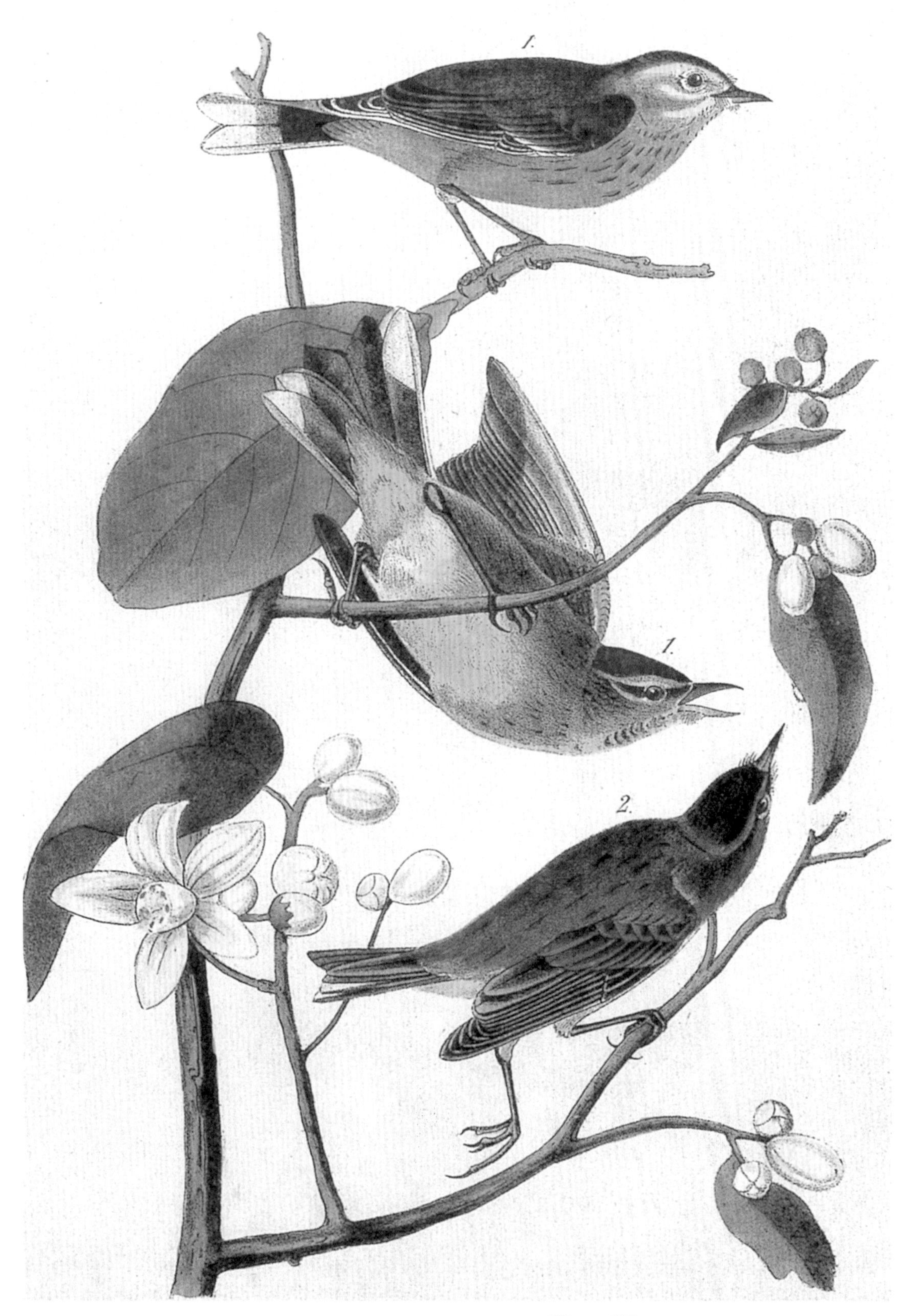

Yellow Red-poll Wood Warbler.

1. Males 2. Young.

Wild Orange Tree.

Drawn from Nature by J.J.Audubon.F.R.S.F.L.S.　　Lith. Printed & Col.d by J. T. Bowen, Philad.a

棕榈林莺（1雄性、2幼鸟）　学名：Dendroica palmarum　英文名：Yellow Red-poll Wood-warbler

Blue yellow-backed Wood-Warbler.

1. Male 2. Female.

Louisiana Flag.

Drawn from Nature by J.J.Audubon,F.R.S.F.L.S.

Lith.d Printed & Col.d by J. T. Bowen, Philad.a

黄背蓝林莺（1雄性、2雌性）

学名：Parula americana 英文名：Blue Yellow-backed Wood-warbler

Pl. 92.

Townsend's Wood Warbler

Male.

Carolina Allspice

Drawn from Nature by J.J.Audubon,F.R.S.F.L.S

Lith^d Printed & Col^d by J. T. Bowen, Philad^a

汤森林莺（雄性）

学名：Dendroica townsendi 英文名：Townsend's Wood-Warbler

Pl. 93.

Hermit Wood-Warbler.

1. Male. 2. Female

Strawberry Tree

Drawn from Nature by J.J.Audubon,F.R.S.F.L.S.　　Lith.d Printed & Col.d by J. T. Bowen, Philad.a

隐士林莺（1雄性、2雌性）　　学名：Dendroica occidentalis　英文名：Hermit Wood-warbler

Black-throated Grey Wood Warbler.

Males.

Drawn from Nature by J.J.Audubon,F.R.S.F.L.S. Lith.d Printed & Col.d by J. T. Bowen, Philad.a

黑喉灰林莺（雄性） 学名：Dendroica nigrescens 英文名：Black-throated Grey Wood-warbler

黑喉蓝林莺（1雄性、2雌性）

学名：Dendroica caerulescens 英文名：Black-throated Blue Wood-warbler

Pl. 96

Black & yellow Wood-Warbler.

1 Male 2 Female 3 Young.

Flowering Raspberry Rubus odoratus.

Drawn from Nature by J.J.Audubon.F.R.S.F.L.S. Lith.d Printed & Col.d by J. T. Bowen Philad.a

黑黄林莺（1雄性、2雌性、3幼鸟） 学名：Dendroica magnolia 英文名：Black & Yellow Wood-warbler

Connecticut Warbler.

1. Male. 2. Female.

Gentiana Saponaria.

Drawn from Nature by J.J.Audubon,F.R.S.F.L.S.

Lith.d Printed & Col.d by J. T. Bowen Philad.a

康涅狄格林莺（1雄性、2雌性）

学名：Oporornis agilis 英文名：Connecticut Warbler

Mourning Ground-Warbler.

Male.

Pheasant's-eye Flos-Adonis.

Drawn from Nature by J.J.Audubon,F.R.S.F.L.S.

Lith.d Printed & Col.d by J. T. Bowen, Philad.a

悲鸣林莺（1雄性、2雌性）

学名：Oporornis tolmiei　英文名：Mourning Ground-warbler

Swainson's Swamp Warbler

Male.

Orange-coloured Azalea. Azalea calendulacea

Drawn from Nature by J.J.Audubon.F.R.S.F.L.S.

Lith.d Printed & Col.d by J. T. Bowen, Philad.a

苦马沼泽林莺（雄性）

学名：Limnothlypis swainsonii 英文名：Swainson's Swamp Warbler

Pl. 105.

Worm-eating Swamp Warbler.

1 Male. 2 Female.

American Poke-weed. Phytolacca decandra.

Drawn from Nature by J.J.Audubon, F.R.S.F.L.S.

Lith.d Printed & Col.d by J. T. Bowen, Philad.a

食虫沼泽林莺（1雄性、2雌性）

学名：Helmitheros vermivorus　英文名：Worm-eating Swamp Warbler

Prothonotary Swamp-Warbler.

1. Male. 2. Female.

Cane Vine.

Drawn from Nature by J.J. Audubon, F.R.S.F.L.S.   Lith.d Printed & Col.d by J. T. Bowen Philad.a

书记湿地林莺（1雄性、2雌性）

学名：Protonataria citrea　英文名：Prothonotary Swamp-warbler

Bachman's Swamp Warbler.

1. Male. 2. Female.

Gordonia pubescens?

Drawn from Nature by J.J.Audubon.F.R.S.F.L.S.　　Lith.d Printed & Col.d by J. T. Bowen, Philad.a

巴赫湿地林莺（1雄性、2雌性）　　学名：Vermivora bachmanii　英文名：Bachman's Swamp Warbler

Tennessee Swamp Warbler.

Male

Ilex laxifolia.

Drawn from Nature by J.J.Audubon,F.R.S.F.L.S. Lith.d Printed & Col.d by J. T. Bowen, Philad.a

田纳西州湿地莺（雄性）

学名：Vermivora peregrina 英文名：Tennessee Swamp Warbler

Blue-winged Yellow Swamp-Warbler.

1. Male. 2. Female.

Cotton Rose. Hibiscus grandiflorus.

Drawn from Nature by J.J.Audubon,F.R.S.F.L.S. Lith.d Printed & Col.d by J.T.Bowen Philad.a

蓝翅黄莺（1雄性、2雌性） 学名：Vermivora pinus 英文名：Blue-winged Yellow Swamp-warbler

Black-and-white Creeping Warbler.

Male

Black Larch Pinus-pendula.

Drawn from Nature by J.J.Audubon,F.R.S.F.L.S.

Lith.d Printed & Col.d by J. T. Bowen, Philad.a

黑白苔莺（雄性）

学名：Mniotilta varia　英文名：Black-and-white Creeping Warbler

Pl. 115

Brown Tree-creeper.

1. Male 2. Female.

Drawn from Nature by J.J.Audubon.F.R.S.F.L.S. Lith.d Printed & Col.d by J. T Bowen Philad.a

褐色爬刺莺（1雄性、2雌性） 学名：Certhia americana 英文名：Brown Tree-Creeper

Rock - Wren.

Adult Female.

Smilacina borealis.

Drawn from Nature by J.J.Audubon,F.R.S.F.L.S

Lith.d Printed & Col.d by J. T. Bowen, Philad.a

岩石鷦鷯（雌性）

学名：Salpinctes obsoletus　英文名：Rock Wren

Great Carolina Wren.

1 Male 2. Female.

Dwarf Buck-eye. Æsculus. Paviæ.

Drawn from Nature by J.J.Audubon,F.R.S.F.L.S. Lith.d Printed & Col.d by J. T. Bowen, Philad.a

卡罗莱纳州鹪鹩（1雄性、2雌性） 学名：Thryothorus ludovicianus 英文名：Great Carolina Wren

Bewicks Wren.

Male.

Iron-wood Tree.

Drawn from Nature by J.J. Audubon, F.R.S.F.L.S.

Lith.d Printed & Col.d by J. T. Bowen, Philad.a

比威克鹪鹩（雄性）

学名：Thryomanes bewickii 英文名：Bewick's Wren

Wood Wren.

Male.

Arbutus. Uva-ursi.

Drawn from Nature by J.J.Audubon.F.R.S.F.L.S.

Lith.d Printed & Col.d by J. T. Bowen. Philad.a

市鹪鹩（雄性）

学名：Troglodytes aedon 英文名：Wood Wren

House Wren

1 Male 2 Female 3 Young

In an old Hat

Drawn from Nature by J.J.Audubon, F.R.S.F.L.S.

Lith.d Printed & Col.d by J.T.Bowen, Philad.a

家鷦鹩（1雄性、2雌性、3幼鸟）

学名：Troglodytes aedon　英文名：House Wren

Pl. 121.

Winter Wren

1. Male 2. Female 3. Young

Drawn from Nature by J.J.Audubon.F.R.S.F.L.S. Lith.d Printed & Col.d by J. T. Bowen Philad.a

冬鷦鷯（1雄性、2雌性、3幼鸟） 学名：Troglodytes troglodytes 英文名：Winter Wren

1. Males. 2 Female & Nest.

泽地鹪鹩（1雄性、2雌性）

学名：Cistothorus palustris 英文名：Marsh Wren

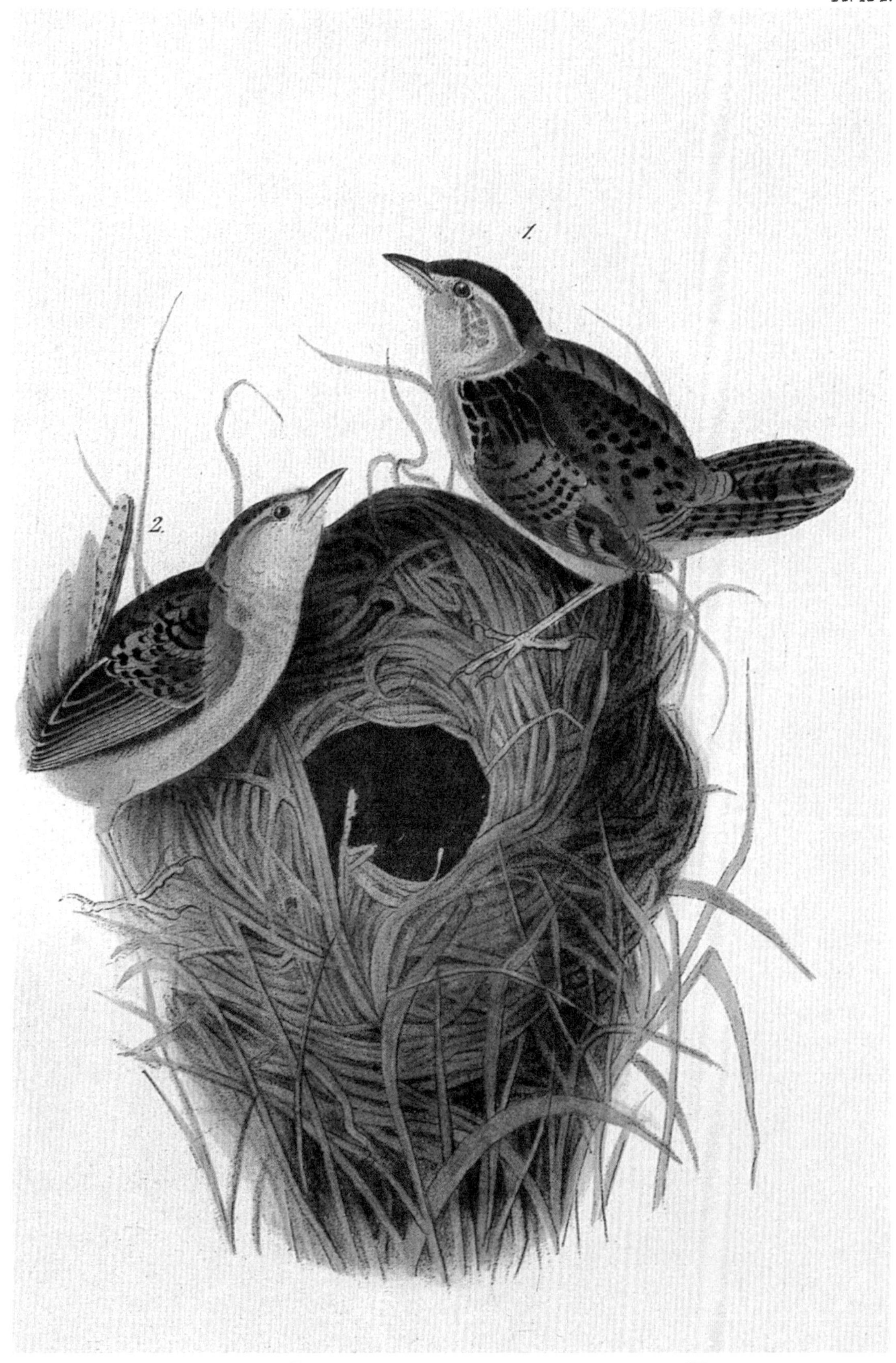

Short-billed Marsh Wren

1. Male. 2. Female and Nest.

Drawn from Nature by J.J. Audubon, F.R.S.F.L.S.

Lith.d Printed & Col.d by J. T. Bowen, Philad.a

莎草鹪鹩（1雄性、2雌性）

学名：Cistothorus platensis 英文名：Short-billed Marsh Wren

Pl. 125

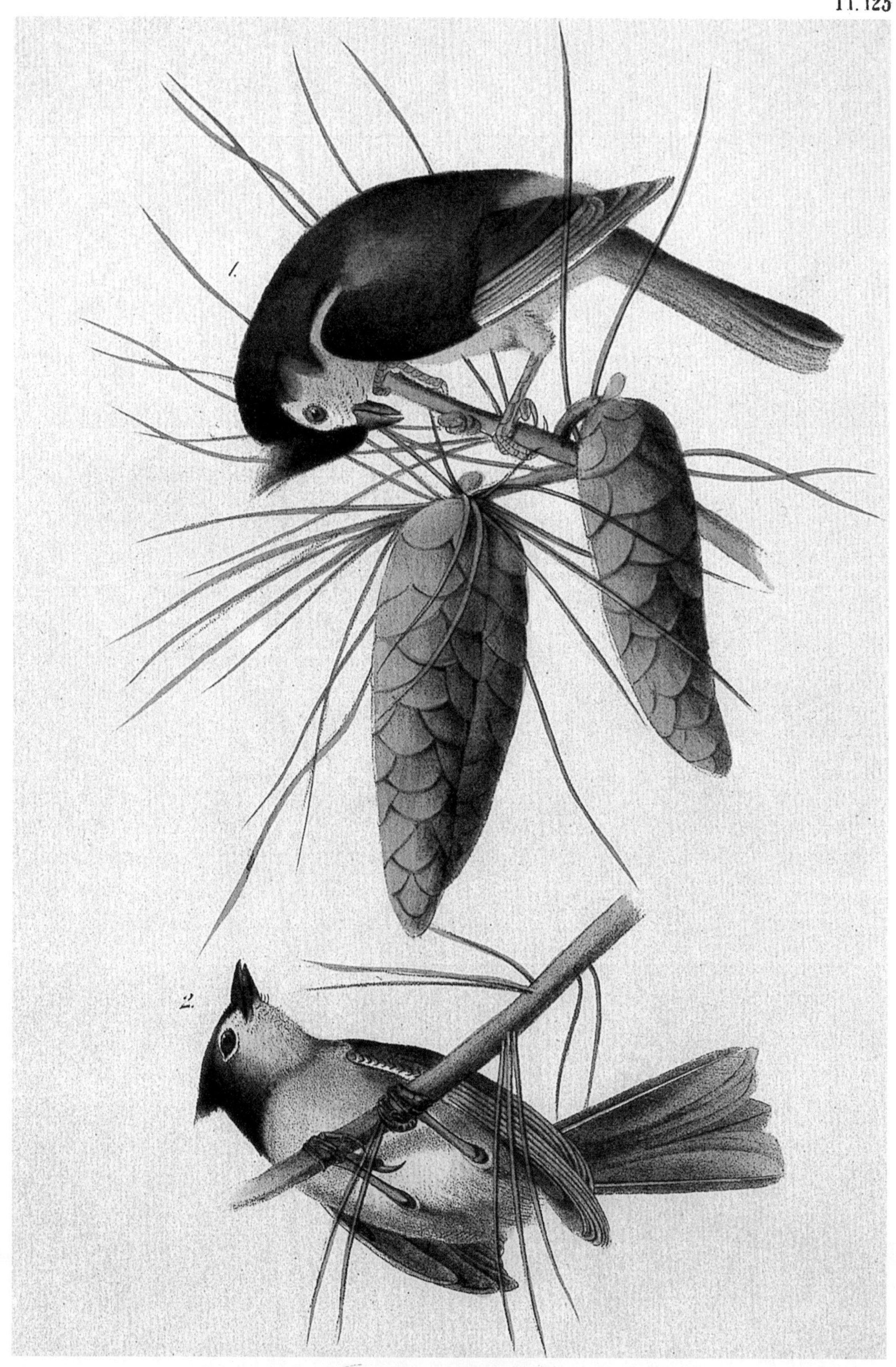

Crested Titmouse.

1. Male. 2. Female.

White Pine. Pinus Strobus.

Drawn from Nature by J.J.Audubon,F.R.S.F.L.S. Lith.d Printed & Col.d by J. T. Bowen, Philad.a

冠顶山雀（1雄性、2雌性）

学名：Parus atricristatus 英文名：Crested Titmouse

Black cap Titmouse

1. Male 2. Female

Plant. Sweet briar

Drawn from Nature by J.J. Audubon, F.R.S.F.L.S. Lith.d Printed & Col.d by J.T. Bowen Philad.a

黑顶山雀（1雄性、2雌性） 学名：Parus atricapillus 英文名：Black Cap Titmouse

Pl. 127.

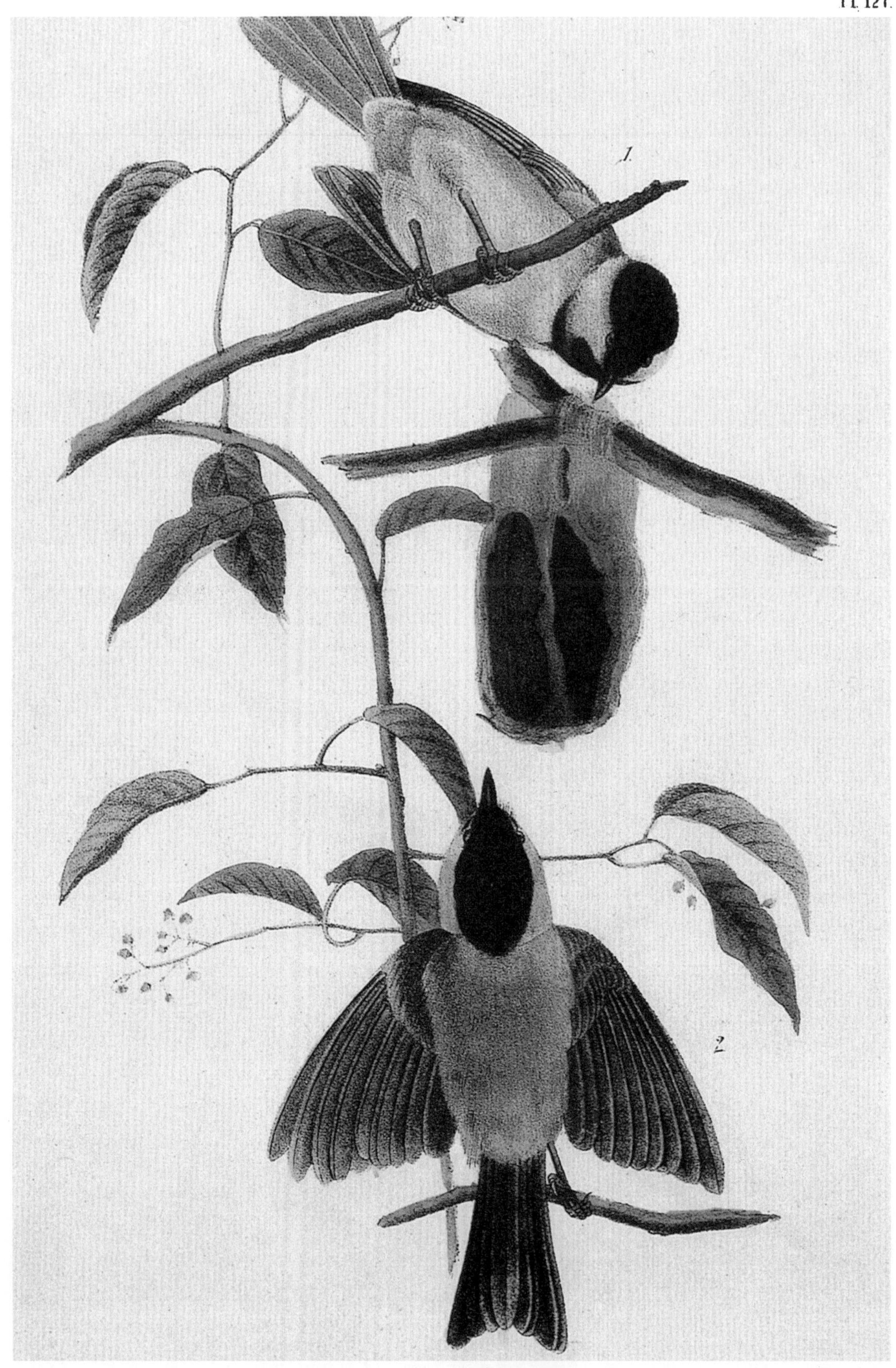

Carolina Titmouse.

1. Male. 2. Female.

Plant. Supple. Jack?

Drawn from Nature by J.J.Audubon F.R.S.F.L.S.　　Lith.d Printed & Col.d by J.T. Bowen, Philad.a

卡罗莱纳州山雀（1雄性、2雌性）　　学名：Parus carolinensis　英文名：Carolina Titmouse

PL. 128

Hudson's Bay Titmouse.

1. Male. 2. Female 3. Young.

Drawn from Nature by J.J.Audubon.F.R.S.F.L.S.

Lith.d Printed & Col.d by J. T. Bowen, Philad.a

北山雀（1雄性、2雌性、3幼鸟）

学名：Parus hudsonicus　英文名：Hudson's Bay Titmouse

Chesnut backed Titmouse.

1. Male. 2. Female.

Drawn from Nature by J.J.Audubon,F.R.S.F.L.S.   Lith.^d Printed & Col.^d by J. T. Bowen, Philad.^a

赤背山雀（1雄性、2雌性）   学名：Parus rufescens 英文名：Chestnut Backed Titmouse

Chesnut-crowned Titmouse.

1. Male. 2. Female and Nest.

Drawn from Nature by J.J.Audubon, F.R.S.F.L.S.　　Lith.d Printed & Col.d by J. T. Bowen, Philad.a

栗冠山雀（1雄性、2雌性）　　学名：Psaltriparus minimus　英文名：Chestnut-crowned Titmouse

Pl. 131.

Cuvier's Kinglet.

Male.

Broad-leaved laurel Kalmia latifolia.

Drawn from Nature by J.J.Audubon,F.R.S.F.L.S.

Lith.d Printed & Col.d by J. T. Bowen, Philad.a

居维叶戴菊鸟（雄性）

学名：Regulus cuvieri 英文名：Cuvier's Kinglet

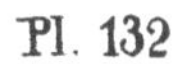

American Golden-crested Kinglet.

1. Male. 2. Female.

Thalia dealbata.

Drawn from Nature by J.J.Audubon.F.R.S.F.L.S. Lith.d Printed & Col.d by J. T Bowen Philad.a

火冠戴菊鸟（1雄性、2雌性） 学名：Regulus satrapa 英文名：American Golden-crested Kinglet

Ruby-crowned Kinglet

1. Male. 2. Female.

Kalmia angustifolia.

Drawn from Nature by J.J.Audubon,F.R.S.F.L.S. Lith.d Printed & Col.d by J. T. Bowen, Philad.a

红冠戴菊鸟（1雄性、2雌性） 学名：Regulus calendula 英文名：Ruby-crowned Kinglet

Common Blue Bird.

1. Male. 2. Female. 3. Young

Great Mullein Verbascum Thapsus

Drawn from Nature by J.J.Audubon, F.R.S.F.L.S. Lith.d Printed & Col.d by J. T. Bowen, Philad.a

知更鸟（1雄性、2雌性） 学名：Sialia sialis 英文名：Common Blue Bird

Pl. 135.

Western Blue Bird.

1. Male. 2. Female.

Drawn from Nature by J.J.Audubon.F.R.S.F.L.S.

Lith. Printed & Col.d by J. T. Bowen, Philad.a

西蓝鸲（1雄性、2雌性）

学名：Sialia mexicana　英文名：Western Blue Bird

Arctic Blue Bird

Male 1. Female 2

Drawn from Nature by J.J.Audubon.F.R.S.F.L.S

Lith. Printed & Col^{d} by J.T.Bowen, Phil

蓝知更鸟（1雄性、2雌性）

学名：Sialia currucoides 英文名：Arctic Blue Bird

北嘲鸟（1雄性、2雌性、3响尾蛇） 学名：Mimus polyglottos 英文名：Common Mocking

Mountain Mocking Bird

Male

Drawn from Nature by J J Audubon, F R S F L S

Lith Printed & Col^d by J T Bowen, Phi

山嘲鸟（雄性）

学名：Oreoscoptes montanus　英文名：Mountain Mocking Bird

Ferruginous Mocking Bird?

Males. 1. 2. 3. Female 4.

Drawn from Nature by J.J.Audubon.F.R.S.F.L.S

Lith.d Printed & Col.d by J. T Bowen. Philad.a

褐嘲鸟（1、2、3雄性、4雌性）

学名：Toxostoma rufum　英文名：Ferruginous Mocking Bird

American Robin, or Migratory Thrush.

1. Male. 2 Female and Young.

Chesnut Oak Quercus prinus.

Drawn from Nature by J.J.Audubon,F.R.S.F.L.S

Lith.d Printed & Col.d by J. T. Bowen Philad.a

迁徙画眉（1雄性、2雌性和幼鸟）

学名：Turdus migratorius　英文名：American Robin or Migratory Thrush

Varied Thrush.

1. Male. 2. Female.

American Mistletoe, Viscum verticillatum.

Drawn from Nature by J.J. Audubon, F.R.S.F.L.S.

Lith.d Printed & Col.d by J. T. Bowen, Philad.a

杂色画眉（1雄性、2雌性）

学名：Ixoreus naevius　英文名：Varied Thrush

Wood Thrush.

1. Male 2. Female.

Common Dogwood.

Drawn from Nature by J.J.Audubon F.R.S.F.L.S.

Lith.d Printed & Col.d by J. T. Bowen Philad.a

画眉（1雄性、2雌性）

学名：Hylocichla mustelina 英文名：Wood Thrush

Pl. 145.

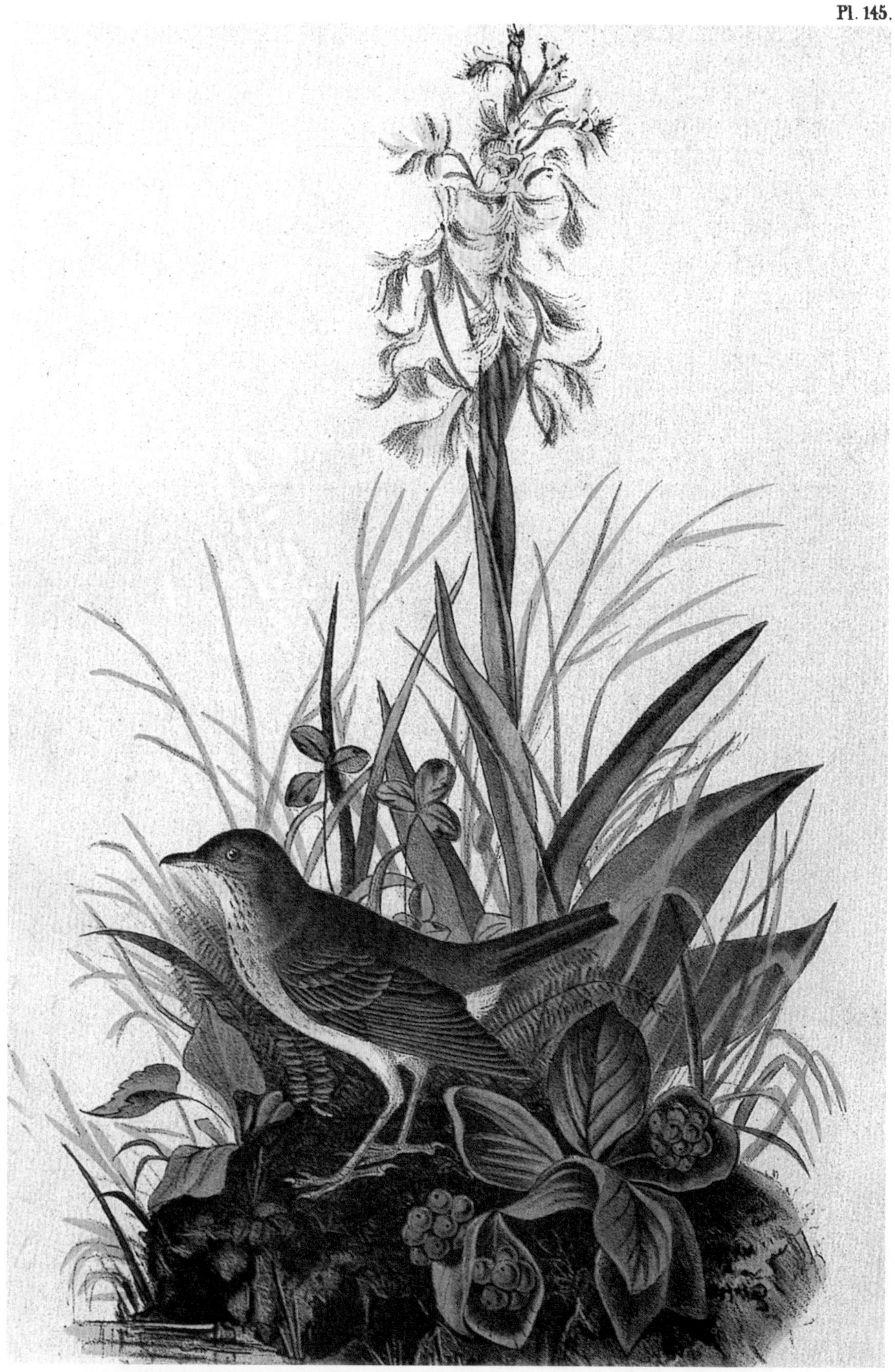

Tawny Thrush

Male,

Habenaria Lacera - Cornus Canadensis

Drawn from Nature by J.J.Audubon, F.R.S.F.L.S.

Lith.d Printed & Col.d by J. T. Bowen, Philad.

茶色画眉（雄性）

学名：Catharus fuscescens　英文名：Tawny Thrush

Hermit Thrush.

1. Male. 2. Female.

Plant Robin Wood.

Drawn from Nature by J.J. Audubon, F.R.S.F.L.S. Lith.d Printed & Col.d by J. T. Bowen, Philad.a

隐居画眉（1雄性、2雌性） 学名：Catharus guttatus 英文名：Hermit Thrush

Drawn from Nature by J.J.Audubon, F.R.S.F.L.S. Lith.d Printed & Col.d by J. T. Bowen, Philad.a

金冠画眉（1雄性、2雌性） 学名：Seiurus aurocapillus 英文名：Golden Crowned Wagtail (Thrush)

Pl 149.

水栖画眉（1雄性、2雌性）

学名：Seiurus motacilla 英文名：Aquatic Wood-wagtail

Pl. 489

Missouri Meadow Lark.

Male.

Drawn from Nature by J.J.Audubon, F.R.S.F.L.S. Lith.d Printed & Col.d by J. T. Bowen, Phila.

密苏里州野云雀（雄性）

学名：Sturnella magna　英文名：Missouri Meadow Lark

Pl. 151.

Shore Lark.

1. Male. Summer Plumage 2. D° Winter 3. Female 4. Young & Nest.

Drawn from Nature by J.J.Audubon, F.R.S.F.L.S.

Lith.d Printed & Col.d by J. T. Bowen, Philad.a

海滨百灵

学名：Eremophila alpestris 英文名：Shore Lark 1雄性夏羽、2雄性冬羽、3雌性、4幼鸟

Pl. 152

Lapland Lark Bunting

1. Male Spring Plumage 2. D° Winter. 3. Female.

Drawn from Nature by J.J.Audubon, F.R.S.F.L.S.

Lith.d Printed & Col.d by J. T. Bowen, Philad.a

拉普兰白颊百灵（1雄性春羽、2雄性冬羽、3雌性）

学名：Calcarius lapponicus 英文名：Lapland Lark Bunting

Pl. 493.

Shattucks Bunting?

Male.

Drawn from Nature by J.J.Audubon,F.R.S.F.L.S.

Lith.d Printed & Col.d by J. T. Bowen, Philad.a

灰雀（雄性）

学名：Spizella pallida　英文名：Shattuck's Bunting

Le Contis Sharp-tailed Bunting.

Male.

Drawn from Nature by J.J.Audubon.F.R.S.F.L.S.　　Lith.d Printed & Col.d by J. T. Bowen. Philad.a

尖尾鹀（雄性）　　学名：Passerherbulus cauda cutus　英文名：Le Conte's Sharp-tailed Bunting

Painted Lark-Bunting.

Male.

Drawn from Nature by J.J Audubon, F.R.S.F.L.S.

Lith.d Printed & Col.d by J. T. Bowen, Philad.a

白颊百灵（雄性）

学名：Calcarius pictus　英文名：Painted-lark Bunting

Chesnut-collared Lark Bunting.

Male

Drawn from Nature by J.J.Audubon, F.R.S.F.L.S.

Lith^d Printed & Col^d by J.T. Bowen, Philad^a

赤颈百灵（雄性）

学名：Calcarius ornatus　英文名：Chestnut-collared Lark Bunting

Snow Lark Bunting.

1. 2. Adult. 3 Young.

Drawn from Nature by J.J. Audubon F.R.S.F.L.S.

Lith. Printed & Colᵈ by J.T. Bowen. Philadᵃ

雪云雀（1、2成鸟、3幼鸟）

学名：Plectrophenax nivalis　英文名：Snow Lark Bunting

Townsend's Bunting

Male.

Drawn from Nature by J.J.Audubon, F.R.S.F.L.S. Lith.d Printed & Col.d by J. T. Bowen, Philad.a

汤森云雀（雄性） 学名：Emberiza townsendi 英文名：Townsend's Bunting

Bay-winged Bunting.

Male.

Prickly Pear Cactus Opuntia.

Drawn by J. J. Audubon, F.R.S.F.L.S.

Lith. & Col. by Bowen & Co. Philada.

湾翼麻雀（雄性） 学名：Pooecetes gramineus 英文名：Bay-winged Bunting

Savannah Bunting.

1. Male 2. Female

Indian Pink-root. Spigelia marilandica.

Drawn from Nature by J.J. Audubon F.R.S.F.L.S.　　Lith.d & Col.d by Bowen & Co Philad.a

草原麻雀（1雄性、2雌性）　　学名：Passerculus sandwichensis　英文名：Savannah Bunting

Clay-coloured Bunting.

Male.

Asclepias tuberosa.

Drawn from Nature by J.J. Audubon, F.R.S.F.L.S.　　Lith.d Printed & Col.d by J. T. Bowen, Philad.a

土色麻雀（雄性）　　学名：Spizella pallida　英文名：Clay-colored Bunting

Yellow-winged Bunting.

Male.

Drawn from Nature by J.J.Audubon,F.R.S.F.L.S.　　　Lith.d Printed & Col.d by J.T.Bowen, Philad.a

黄翼麻雀（雄性）　　　学名：Ammodramus savannarum　英文名：Yellow-winged Bunting

Henslow's Bunting
Male.
Indian Pink root or Worm grass
Spigelia Marilandica
Phlox aristata.

Drawn from Nature by J.J.Audubon,F.R.S.F.L.S.　　Lith.d Printed & Col.d by J. T. Bowen, Philad.a

亨斯麻雀（雄性）　　学名：Ammodramus henslowii　英文名：Henslow's Bunting

Pl. 164.

Field Bunting.

Male

Calopogon pulchellus. Brown
Dwarf Huckle-berry Vaccinium tenellum

Drawn from Nature by J.J.Audubon, F.R.S.F.L.S.

Lith.d Printed & Col.d by J.T. Bowen Philad.a

田间麻雀（雄性）

学名：Spizella pusilla　英文名：Field Bunting

Pl. 165

Chipping Bunting.

Male.

Black locust or False Acacia. Robinia pseudacacia.

Drawn from Nature by J.J Audubon, F.R.S.F.L.S.

碎屑麻雀（雄性） 学名：Spizella passerina 英文名：Chipping Bunting

Pl. 168.

Oregon Snow-Bird.

1 Male. 2 Female.

Rosa Laevigata?

Drawn from Nature by J.J.Audubon, F.R.S.F.L.S.　　Lith.d Printed & Col.d by J. T. Bowen, Philad.a

灯心草雀（1雄性、2雌性）　　学名：Junco hyemalis　英文名：Oregon Snow Bird

Painted Bunting

1.2.3. Males in different States of Plumage. 4. Female.

Chicasaw Wild Plum.

Drawn from Nature by J.J.Audubon.F.R.S.F.L.S.

Lith.d Printed & Col.d by J. T. Bowen, Philad.a

五彩雀（1、2、3雄性、4雌性）

学名：Passerina ciris　英文名：Painted Bunting

Pine Linnet

1 Male 2 Female

Black Larch

Drawn from Nature by J.J. Audubon, F.R.S.F.L.S.

Lith. Printed & Col.d by J.T. Bowen Philad.

青雀（1、2、3雄性、4雌性）

学名：Passerina cyanea　英文名：Jndigo Bunting

Lazuli Finch.

1. Male 2. Female.

Wild Spanish Coffee.

Drawn from Nature by J.J.Audubon,F.R.S.F.L.S.

Lith.d Printed & Col.d by J. T. Bowen, Philad.a

天青石雀（1雄性、2雌性）

学名：Passerina amoena 英文名：Lazuli Finch

Sea-side Finch.

1. Male 2. Female.

Carolina Rose.

Drawn from Nature by J.J.Audubon.F.R.S.F.L.S. | Lith.d Printed & Col.d by J. T. Bowen, Philad.a

海滨雀（1雄性、2雌性） 学名：Ammodramus maritimus 英文名：Sea-side Finch

Sharp-tailed Finch.

1. Male. 2. Female & Nest.

Drawn from Nature by J.J.Audubon, F.R.S.F.L.S.　　Lith.d Printed & Col.d by J. T. Bowen, Philad.a

尖尾雀（1雄性、2雌性）　　学名：Ammodramus caudacutus　英文名：Sharp-tailed Finch

Swamp Sparrow

Male.

May-apple

Drawn from Nature by J.J.Audubon,F.R.S.F.L.S. Lith.d Printed & Col.d by J.T.Bowen. Philad.a

湿地麻雀（雄性） 学名：Melospiza georgiana 英文名：Swamp Sparrow

Bachman's Pinewood Finch

Male.

Pinckneya pubescens.

Drawn from Nature by J.J.Audubon,F.R.S.F.L.S.

Lith.d Printed & Col.d by J. T. Bowen, Philad.a

巴赫松雀（雄性）

学名：Aimophila aestivalis　英文名：Bachman's Pinewood Finch

1. Male. 2 Female.

1. Dwarf Cornel 2 Cloudberry 3. Glaucous Kalmia.

Drawn from Nature by J.J.Audubon, F.R.S.F.L.S.

Lith.d Printed & Col.d by J. T. Bowen, Philad.

林肯松雀（1雄性、2雌性）

学名：Melospiza lincolnii 英文名：Lincoln's Pinewood-Finch

Mealy Redpoll Linnet.

Male.

Drawn from Nature by J.J.Audubon F.R.S.F.L.S

Lith.d Printed & Col.d by J. T. Bowen, Philad.a

粉末金翅鸟（雄性）

学名：Carduelis hornemanni　英文名：Mealy Redpoll Linnet

Pl. 179.

Lesser Redpoll Linnet

1 Male. 2 Female.

Drawn from Nature by J.J. Audubon F.R.S.F.L.S.　　Lith^d Printed & Col^d by J. T. Bowen Philad^a

小红雀（1雄性、2雌性）　　学名：Carduelis flammea　英文名：Lesser Redpoll Linnet

Pl. 180.

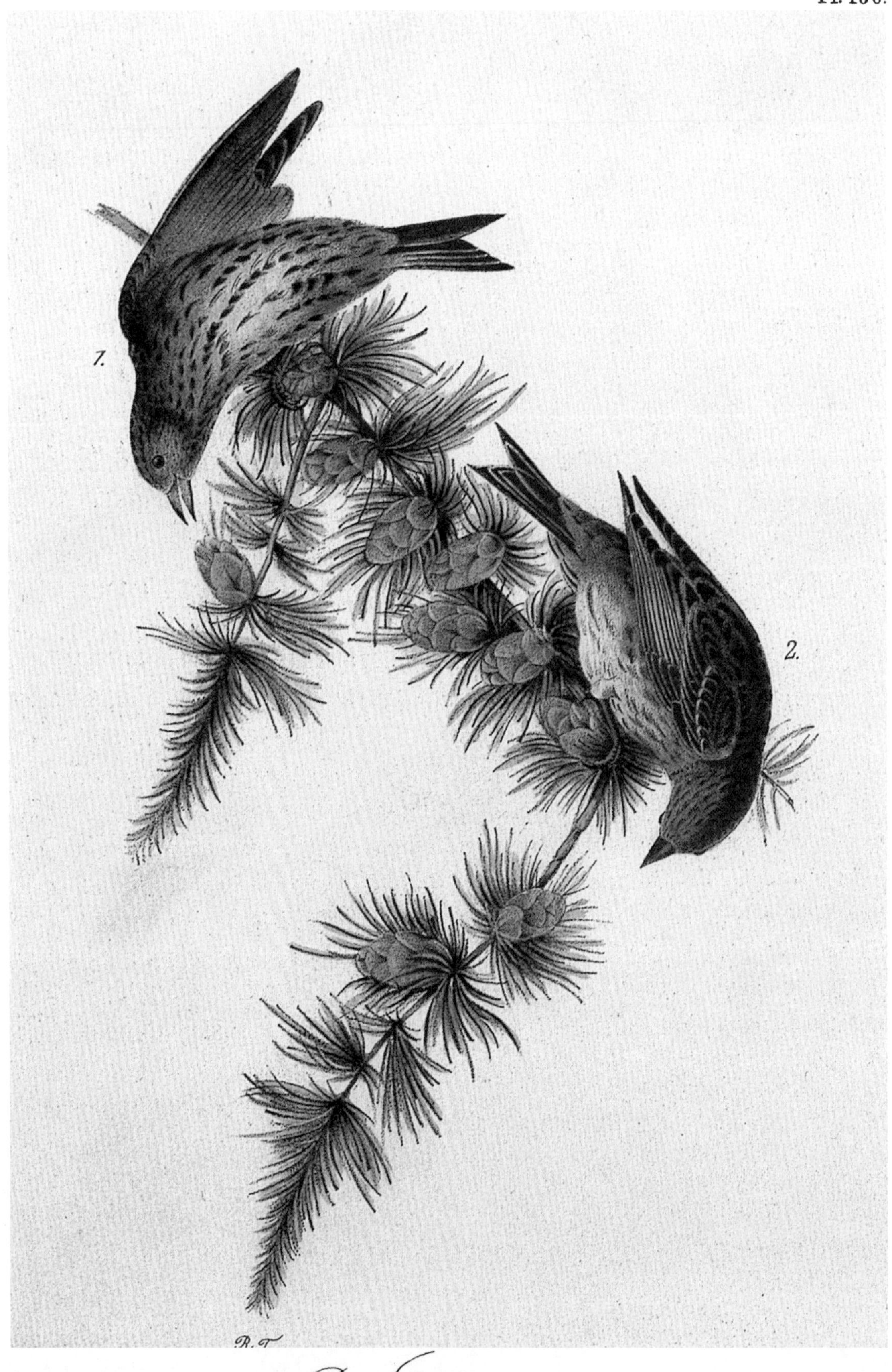

松红雀（1雄性、2雌性）　学名：Carduelis pinus　英文名：Pine Linnet

Pl. 181.

American Goldfinch

1. Male. 2. Female.

Common Thistle.

Drawn from Nature by J.J.Audubon, F.R.S.F.L.S

Lith^d Printed & Col^d by J. T. Bowen, Philad^a

美洲金翅鸟（1雄性、2雌性）

学名：Carduelis tristis　英文名：American Goldfinch

Black-headed Goldfinch

Male.

Drawn from Nature by J.J.Audubon, F.R.S.F.L.S.

Lith.d Printed & Col.d by J. T. Bowen, Philad.a

黑头金翅鸟（雄性）

学名：Carduelis magellanica　英文名：Black-headed Goldfinch

Pl. 184.

Yarrell's Goldfinch

1. Male. 2. Female

Drawn from Nature by J.J. Audubon, F.R.S.F.L.S.　　Lith.d Printed & Col.d by J. T. Bowen, Philad.a

小金翅鸟（1雄性、2雌性）　　学名：Carduelis psaltria　英文名：Yarrell's Goldfinch

Stanley Goldfinch

Drawn from Nature by J.J.Audubon,F.R.S.F.L.S. Lith.d Printed & Col.d by J. T. Bowen, Philad.a

黑衣金翅鸟

学名：Carduelis barbatus 英文名：Stanley Goldfinch

Pl. 194

Arctic Ground Finch.

1 Male 2 Female.

Drawn from Nature by J.J.Audubon F.R.S.F.L.S.

Lith.d Printed & Col.d by J. T. Bowen, Philad.a

赤边红眼雀

学名：Pipilo erythrophthalmus 英文名：Arctic Ground Finch 1雄性、2雌性

Fox-coloured Finch

1. Male 2. Female.

Drawn from Nature by J.J.Audubon F.R.S.F.L.S.

Lith.d Printed & Col.d by J. T. Bowen, Philad.a

红斑雀（1雄性、2雌性）

学名：Passerella ilica 英文名：Fox-colored Finch

Harris' Finch.

1. Adult Male 2. Young Female.

Drawn from Nature by J.J.Audubon.F.R.S.F.L.S. Lith^d Printed & Col^d by J.T.Bowen, Philad^a

哈里斯雀（1雄性、2雌性幼鸟） 学名：Zonotrichia querula 英文名：Harris' Finch

Brown Finch

Female.

Drawn from Nature by J. J. Audubon, F.R.S.F.L.S.

Lith.d Printed & Col.d by J. T. Bowen, Philad.a

褐色燕雀（雌性）

学名：Carduelis flammea 英文名：Brown Finch

Song Finch.

1. Male. 2 Female.

Huckle-berry or Blue tangled Vaccinium frondosum.

Drawn from Nature by J.J.Audubon F.R.S.F.L.S

Lith.d Printed & Col.d by J. T. Bowen, Philad.a

善歌雀（1雄性、2雌性）

学名：Melospiza melodia　英文名：Song Finch

Pl. 190

Morton's Finch.

Male.

Drawn from Nature by J.J. Audubon, F.R.S.F.L.S. Lith. Printed & Col.d by J.T. Bowen, Philad.a

莫顿燕雀（雄性） 学名：Zonotrichia capensis 英文名：Morton's Finch

White-throated Finch.

1. Male. 2. Female.

Common Dogwood.

Drawn from Nature by J.J.Audubon, F.R.S.F.L.S.

Lith.d Printed & Col.d by J. T. Bowen, Philad.a

白喉燕雀（1雄性、2雌性）

学名：Zonotrichia albicollis 英文名：White-throated Finch

White-crowned Finch

1 Male. 2 Female.

Wild Summer Grape.

Drawn from Nature by J.J.Audubon,F.R.S.F.L.S. Lith.d Printed & Col.d by J.T. Bowen Philad.a

白冠雀（1雄性、2雌性） 学名：Zonotrichia leucophrys 英文名：White-crowned Finch

Black-and-yellow-crowned Finch.

Drawn from Nature by J.J.Audubon, F.R.S.F.L.S. Lith.d Printed & Col.d by J. T. Bowen, Philad.a

金冠雀 学名：Zonotrichia atricapilla 英文名：Black-and-Yellow-crowned Finch

Towhe Ground Finch

1. Male. 2. Female.

Common Blackberry

Drawn from Nature by J.J. Audubon, F.R.S.F.L.S.

Lith.d Printed & Col.d by J.T. Bowen Philad.a

褐色红眼雀（1雄性、2雌性）

学名：Pipilo erythrophthalmus 英文名：Towhe Ground Finch

Crested Purple Finch

1. Males. 2. Female

Red Larch. Larix Americana.

Drawn from Nature by J.J.Audubon F.R.S.F.L.S.   Lith.^d Printed & Col.^d by J. T. Bowen, Philad.^a

紫雀（1雄性、2雌性）   学名：Carpodacus purpureus　英文名：Crested Purple Finch

Crimson-fronted Purple Finch.

Male.

Drawn from Nature by J.J.Audubon,F.R.S.F.L.S.

Lith.d Printed & Col.d by J. T. Bowen. Philad.a

家雀（雄性）　学名：Carpodacus mexicanus　英文名：Crimson-fronted Purple-finch

Pl. 198.

Grey-crowned Purple Finch.

Male.

Stokesia cyanea

Drawn from Nature by J.J.Audubon, F.R.S.F.L.S. Lith^d Printed & Col^d by J. T. Bowen Philad^a.

玫瑰雀（雄性） 学名：Leucosticte arctoa 英文名：Grey-crowned Purple-finch

Pl. 200.

Common Crossbill.

1. Males. 2. Females.

Drawn from Nature by J.J.Audubon, F.R.S.F.L.S. Lith.d Printed & Col.d by J. T. Bowen, Philad.a

交喙鸟（1雄性、2雌性） 学名：Loxia curvirostra 英文名：Common Crossbill

White-winged Crossbill.

1. Males. 2. Female.

Drawn from Nature by J.J.Audubon, F.R.S.F.L.S.　　Lith^d Printed & Col^d by J. T. Bowen, Philad^a

白翼交喙鸟（1雄性、2雌性）　　学名：Loxia leucoptera　英文名：White-winged Crossbill

Pl. 202.

Prairie Lark Finch.

1 Male. 2 Female.

Drawn from Nature by J.J. Audubon, F.R.S.F.L.S.

Lith.d Printed & Col.d by J. T. Bowen, Philad.a

草原雀（1雄性、2雌性）

学名：Calamospiza melanocorys　英文名：Prairie Lark-finch

Common Cardinal Grosbeak

1. Male. 2. Female

Wild Almond. Prunus caroliniana.

Drawn from Nature by J.J.Audubon,F.R.S.F.L. Lith.d Printed & Col.d by J.T.Bowen, Philad.a

红衣主教蜡嘴鸟（1雄性、2雌性） 学名：Cardinalis cardinalis 英文名：Common Cardinal Grosbeak

Blue Song Grosbeak

1. Male 2. Female 3 Young.

Drawn from Nature by J.J Audubon, F.R.S.F.L.S.   Lith.d Printed & Col.d by J. T. Bowen, Philad.a

蓝调蜡嘴鸟（1雄性、2雌性、3幼鸟）   学名：Guiraca caerulea　英文名：Blue Song Grosbeak

Rose-breasted Song-Grosbeak

1. Males. 2. Female. 3. Young Male.

Ground Hemlock Taxus canadensis.

Drawn from Nature by J.J. Audubon, F.R.S.F.L.S.

Lith.d Printed & Col.d by J. T. Bowen, Philad.a

玫瑰胸蜡嘴鸟（1雄性、2雌性、3雄性幼鸟）

学名：Pheucticus ludovicianus　英文名：Rose-breasted Song Grosbeak

Black-headed Song-Grosbeak.

1. Male 2. Female.

Drawn from Nature by J.J.Audubon,F.R.S.F.L.S. — Lith.d Printed & Col.d by J. T. Bowen, Philad.a

黑头蜡嘴鸟（1雄性、2雌性） 学名：Pheucticus melanocephalus 英文名：Black-headed Song Grosbeak

Evening Grosbeak.

1. Male. 2. Female 3. Young Male.

Drawn from Nature by J.J.Audubon, F.R.S.F.L.S. Lith.d Printed & Col.d by J. T. Bowen, Philad.a

傍晚蜡嘴鸟（1雄性、2雌性、3雄性幼鸟） 学名：Coccothraustes vespertinus 英文名：Evening Grosbeak

Summer Red-bird

1. Male. 2. Female. 3. Young Male.

Wild Muscadine. Vitis rotundifolia. Mich

Drawn from Nature by J.J. Audubon, F.R.S.F.L.S.

Lith.d Printed & Col.d by J. T. Bowen, Philad.a

夏季红雀（1雄性、2雌性、3雄性幼鸟）

学名：Piranga rubra 英文名：Summer Red-bird

Scarlet Tanager

1. Male. 2. Female.

Drawn from Nature by J.J.Audubon, F.R.S.F.L.S.

Lith.d Printed & Col.d by J. T. Bowen, Philad.a

红唐纳雀（1雄性、2雌性）

学名：Piranga olivacea　英文名：Scarlet Tanager

Louisiana Tanager

1. Males 2. Female

Drawn from Nature by J.J Audubon, F.R.S.F.L.S.

Lith.d Printed & Col.d by J. T. Bowen, Philad.a

路易斯安那唐纳雀（1雄性、2雌性）

学名：Piranga ludoviciana 英文名：Louisiana Tanager

Pl. 214.

Red-and-white-shouldered Marsh Blackbird.

Male.

Drawn from Nature by J.J.Audubon,F.R.S.F.L.S.

Lith.d Printed & Col.d by J. T. Bowen, Philad.a

红翼湿地黑鸟

学名：Agelaius phoeniceus 英文名：Red-and-White-shouldered Marsh Blackbird

红翼八哥（1雄性、2雄性幼鸟、3雌性）　学名：Onychognathus morio　英文名：Red-winged Starling

Pl. 499.

Common Troupial.

Male.

Drawn from Nature by J.J.Audubon, F.R.S.F.L.S.

Lith.d Printed & Col.d by J. T. Bowen, Philad.a

黄鹂（雄性）

学名：Icterus icterus　英文名：Common Troupial

Pl 217

Baltimore Oriole, or Hang-nest.

1. Male Adult 2. Young Male 3. Female.

Tulip Tree.

Drawn from Nature by J.J.Audubon, F.R.S.F.L.S. Lith.d Printed & Col.d by J. T. Bowen, Philad.a

果园金莺（1雄性、2雄性幼鸟、3雌性） 学名：Icterus spurius 英文名：Baltimore Oriole or Hang-nest

Bullock's Troopial.

1. Male Adult. 2 Young Male. 3. Female

Caprifolium flavum.

Drawn from Nature by J.J. Audubon, F.R.S.F.L.S.

Lith.d Printed & Col.d by J. T. Bowen, Philad.a

北金莺（1雄性、2雄性幼鸟、3雌性）

学名：Icterus galbula　英文名：Bullock's Troopial

Boat-tailed Grackle.

1. Male 2. Female.

Live Oak.

Drawn from Nature by J.J. Audubon, F.R.S.F.L.S. Lith.d Printed & Col.d by J. T. Bowen, Philad.a

船尾白头翁（1雄性、2雌性） 学名：Quiscalus major 英文名：Boat-tailed Grackl

Common, or Purple Crow-Blackbird.

1. Male 2. Female

Maize or Indian Corn

Drawn from Nature by J.J.Audubon,F.R.S.F.L.S.　　Lith^d. Printed & Col^d. by J. T. Bowen, Philad^a.

紫乌鸦（1雄性、2雌性）　　学名：Quiscalus quiscula　英文名：Common or Purple Crow Blackbird

Pl. 222.

Rusty Crow Blackbird.

1 Male. 2 Female. 3 Young.

Black Haw.

Drawn from Nature by J.J.Audubon, F.R.S.F.L.S.　　Lith.d Printed & Col.d by J. T. Bowen, Philad.a

褐山鸟（1雄性、2雌性、3幼鸟）　　学名：Euphagus carolinus　英文名：Rusty Crow Blackbird

Meadow Starling or Meadow Lark

1. Males. 2. Female and Nest.

Yellow flowered Gerardia.

Drawn from Nature by J.J.Audubon, F.R.S.F.L.S. Lith.d Printed & Col.d by J. T. Bowen, Philad.a

草地鹨（1雄性、2雌性） 学名：Sturnella magna 英文名：Meadow Starling or Meadow Lark

Common American Crow.

Male.

Black Walnut.

Drawn from Nature by J.J.Audubon.F.R.S.F.L.S. Lith.d Printed & Col.d by J. T. Bowen. Philad.a

小嘴乌鸦（雄性） 学名：Corvus brachyrhynchos 英文名：Common American Crow

Fish Crow.

1. Male. 2. Female.

Honey Locust.

Drawn from Nature by J.J.Audubon, F.R.S.F.L.S.　　Lith.d Printed & Col.d by J. T. Bowen, Philad.a

鱼鸦（1雄性、2雌性）　　学名：Corvus ossifragus　英文名：Fish Crow

Pl. 227.

Common Magpie

1. Male 2. Female

Drawn from Nature by J.J.Audubon F.R.S.F.L.S.

Lith.d Printed & Col.d by J.T. Bowen Philad.a

喜鹊（1雄性、2雌性）

学名：Pica pica　英文名：Common Magpie

Yellow billed Magpie

Male

Plantanus

Drawn from Nature by J.J.Audubon.F.R.S.F.L.S　　Lith^d Printed & Col^d by J. T. Bowen Philad^a

黄翼鹊（雄性）　　学名：Pica nuttalli　英文名：Yellow-billed Magpie

Columbia Magpie or Jay.

Males.

Drawn from Nature by J.J.Audubon,F.R.S.F.L.S.

Lith.d Printed & Col.d by J. T. Bowen Philad.a

哥伦比亚鹊（雄性）

学名：Calocitta colliei 英文名：Columbia Magpie or Jay

Drawn from Nature by J.J.Audubon,F.R.S.F.L.S

Lith.d Printed & Col.d by J. T. Bowen, Philad.a

蓝冠鸦（1雄性、2、3雌性）

学名：Cyanocitta cristata 英文名：Blue Jay

Pl. 233.

Florida Jay.

1 Male. 2 Female.

Persimon tree Diospyros Virginiana.

Drawn from Nature by J.J.Audubon F.R.S.F.L.S.

Lith.d Printed

佛罗里达鹊（1雄性、2雌性）

学名：Aphelocoma coerulescens 英文名：Florida Jay

Canada Jay.

1. Male. 2. Female 3 Young.

White Oak. Quercus alba.

Drawn from Nature by J.J.Audubon.F.R.S.F.L.S.

Lith.d Printed & Col.d by J. T. Bowen Philad.a

灰鹊（1雄性、2雌性、3幼鸟）

学名：Perisoreus canadensis　英文名：Canada Jay

Clarke's Nutcracker.

1. Male. 2. Female.

Drawn from Nature by J.J.Audubon F.R.S.F.L.S. Lith.d Printed & Col.d by J.T.Bowen Philad.a

克拉克灰鸟（1雄性、2雌性） 学名：Clark's Nutcracker 英文名：Clarke's Nutcracker

Pl 236.

Great American Shrike.

1. Male 2. Female 3. Young.

Crataegus Apiifolia.

Drawn from Nature by J.J.Audubon, F.R.S.F.L.S. Lith.^d Printed & Col.^d by J. T. Bowen, Philad.^a

美洲伯劳鸟（1雄性、2雌性、3幼鸟） 学名：Lanius excubitor 英文名：Great American Shrike

Loggerhead Shrike.

1 Male 2 Female.

Greenbriar or Round-leaved Smilax Smilax Rotundifolia.

Drawn from Nature by J.J.Audubon,F.R.S.F.L.S. Lith.d Printed & Col.d by J.T.Bowen, Philad.a

傻子伯劳鸟（1雄性、2雌性） 学名：Lanius ludovicianus 英文名：Loggerhead Shrike

Bell's Vireo.

Male

Rattle-snake Root.

Drawn from Nature by J.J.Audubon,F.R.S.F.L.S.

Lith^d Printed & Col^d by J. T. Bowen, Philad^a

铃雀（雄性）

学名：Vireo bellii 英文名：Bell's Vireo

Yellow-throated Vireo or Greenlet.

Male

Swamp Snowball Hydrangea quercifolia.

Drawn from Nature by J.J.Audubon.F.R.S.F.L.S. Lith.d Printed & Col.d by J.T. Bowen, Philad.a

黄喉雀（雄性） 学名：Vireo flavifrons 英文名：Yellow-throated Vireo or Greenlet

Pl 239

Solitary Vireo or Greenlet

1. Male. 2. Female.

American Cane. Miegia macrosperma.

Drawn from Nature by J.J.Audubon F.R.S.F.L.S.　　Lith.d Printed & Col.d by J. T. Bowen Phila.d

孤独雀（1雄性、2雌性）　　学名：Vireo solitarius　英文名：Solitary Vireo or Greenlet

Pl. 240.

White-eyed Vireo or Greenlet.

Male.

Pride of China, or bead tree. Melia Azedarach.

Drawn from Nature by J.J.Audubon, F.R.S.F.L.S. Lith.d Printed & Col.d by J.T. Bowen Philad.a

白眼雀（雄性） 学名：Vireo griseus 英文名：White-eyed Vireo or Greenlet

Bartrams Vireo or Greenlet.

Male.
Ipomea.

Drawn from Nature by J.J.Audubon,F.R.S.F.L.S.　　Lith.d Printed & Col.d by J. T. Bowen, Philad.a

绿雀（雄性）　　学名：Vireo olivaceus　英文名：Bartram's Vireo or Greenlet

Red-eyed Vireo or Greenlet

Male

Honey locust

Drawn from Nature by J.J.Audubon.F.R.S.F.L.S. Lith.d Printed & Col.d by J. T. Bowen Philad.a

红眼绿雀（雄性） 学名：Vireo olivaceus 英文名：Red-eyed Vireo or Greenlet

Black throated Wax-wing
or Bohemian Chatterer.

1. Male. 2. Female.
Canadian Service Tree

Drawn from Nature by J.J.Audubon, F.R.S.F.L.S.   Lith.d Printed & Col.d by J. T. Bowen, Philad.a

波希米亚朱缘蜡翅鸟（1雄性、2雌性）   学名：Bombycilla garrulus   英文名：Black-throated Wax-wing or Bohemian Chatterer

White-breasted Nuthatch.

1. Male. 2. & 3. Female.

Drawn from Nature by J.J.Audubon.F.R.S.F.L.S.　　Lith.d Printed & Col.d by J. T. Bowen, Philad.a

白胸五子雀（1雄性，2、3雌性）　　学名：Sitta carolinensis　英文名：White-breasted Nuthatch

Red-bellied Nuthatch

1 Male. 2 Female.

Drawn from Nature by J.J.Audubon.F.R.S.F.L.S

Lith.d Printed & Col.d by J.T Bowen, Philad.a

红腹五子雀（1雄性、2雌性）

学名：Sitta canadensis　英文名：Red-bellied Nuthatch

Pl. 249.

Brown headed Nuthatch.

1. Male. 2. Female

Drawn from Nature by J.J. Audubon, F.R.S.F.L.S — Lith.d Printed & Col.d by J.T. Bowen Philad.a

褐头五子雀（1雄性、2雌性） 学名：Sitta pusilla 英文名：Brown-headed Nuthatch

加州五子雀　　学名：Sitta pygmea　英文名：Californian Nuthatch

Mango Humming bird.

1. 2. Males. 3. Female.

Bignonia grandifolia.

Drawn from Nature by J.J.Audubon, F.R.S.F.L.S.　　Lith.d Printed & Col.d by J. T. Bowen, Philad.a

黑喉蜂鸟（1、2雄性、3雌性）　学名：Trochilus mango　英文名：Mango Humming bird

Anna Humming bird.

1. 2. Males. 3 Female.

Hibiscus Virginicus.

Drawn from Nature by J.J.Audubon,F.R.S.F.L.S.

Lith.d Printed & Col.d by J. T. Bowen, Philad.a

安娜蜂鸟（1、2雄性，3雌性）

学名：Calypte anna　英文名：Anna Humming bird

Pl 254

Ruff-necked Humming bird.

1. 2. Males 3. Female.

Cleome heptaphylla.

Drawn from Nature by J.J.Audubon.F.R.S.F.L.S.

Lith.d Printed & Col.d by J. T Bowen Philad.a

红褐色蜂鸟（1、2雄性、3雌性）　　学名：Selasphorus rufus　英文名：Ruff-necked Humming bird

Ivory-billed Woodpecker.

1. Male. 2. Female.

Drawn from Nature by J.J.Audubon F.R.S.F.L.S. Lith.d Printed & Col.d by J.T. Bowen Philad.a

象牙喙啄木鸟（1雄性、2雌性） 学名：Campephilus principalis 英文名：Ivory-billed Woodpecker

北美黑啄市鸟（1雄性、2雌性、3、4雄性幼鸟） 学名：Dryocopus pileatus 英文名：Pileated Woodpecker

Pl. 260

1. Male. 2. Female.

Drawn from Nature by J.J.Audubon, F.R.S.F.L.S.

Lith.d Printed & Col.d by J. T. Bowen, Philad.a

玛利亚的啄木鸟（1雄性、2雌性）

学名：Picoides villosus　英文名：Maria's Woodpecker

Downy Woodpecker.

1. Male. 2. Female.

Drawn from Nature by J.J. Audubon, F.R.S.F.L.S.

Lith^d Printed & Col^d by J.T. Bowen Philad^a

绒啄市鸟（1雄性、2雌性）

学名：Picoides pubescens　英文名：Downy Woodpecker

Audubons Woodpecker.

Male.

Drawn from Nature by J.J.Audubon, F.R.S.F.L.S.　　Lith.d Printed & Col.d by J. T. Bowen, Philad.a

奥杜邦啄木鸟（雄性）　　学名：Picoides borealis　英文名：Audubon's Woodpecker

Red-breasted Woodpecker.

1. Male. 2. Female.

Drawn from Nature by J.J.Audubon, F.R.S.F.L.S.

Lith.d Printed & Col.d by J. T. Bowen, Philad.a

红胸啄木鸟（1雄性、2雌性）

学名：Sphyrapicus ruber　英文名：Red-breasted Woodpecker

Yellow-bellied Woodpecker.

1. Male. 2. Female.

Prunus Caroliniana.

Drawn from Nature by J.J. Audubon, FRSFLS

Lith.d Printed & Col.d by J.T. Bowen, Philad.a

黄腹啄木鸟（1雄性、2雌性）

学名：Sphyrapicus varius　英文名：Yellow-bellied Woodpecker

1. 2. Males. 3. Female.

Drawn from Nature by J.J.Audubon, F.R.S.F.L.S. Lithd. Printed & Cold. by J. T. Bowen, Philada.

黑背啄木鸟（1、2雄性、3雌性）　学名：Picoides arcticus　英文名：Arctic three-toed Woodpecker

Missouri Red-moustached Woodpecker.

Male.

Drawn from Nature by J.J.Audubon F.R.S.F.L.S

Lith.d Printed & Col.d by J. T. Bowen, Philad.a

红须啄市鸟（雄性）

学名：hybrid-Colaptes auratus x C. cafer　英文名：Missouri Red-moustached Woodpecker

Red-bellied Woodpecker.

1. Male. 2. Female.

Drawn from Nature by J.J. Audubon, F.R.S.F.L.S. Lith.d Printed & Col.d by J. T. Bowen, Philad.a

红腹啄木鸟（1雄性、2雌性） 学名：Melanerpes carolinus 英文名：Red-bellied Woodpecker

Drawn from Nature by J.J.Audubon,F.R.S.F.L.S. Lith^d Printed & Col^d by J. T. Bowen, Philad^a

红头啄木鸟（1雄性、2雌性、3幼鸟） 学名：Melanerpes erythrocephalus 英文名：Red-headed Woodpecker

Lewis' Woodpecker.

1. Male 2. Female

Drawn from Nature by J.J. Audubon, F.R.S.F.L.S. Lith.d Printed & Col.d by J. T. Bowen Philad.a

重爪啄木鸟（1雄性、2雌性） 学名：Melanerpes lewis 英文名：Lewis' Woodpecker

Pl 273

Golden-winged Woodpecker.

1. Male. 2. Females.

Drawn from Nature by J.J.Audubon, F.R.S.F.L.S. 　 Lith.d Printed & Col.d by J. T. Bowen, Philad.a

金翅啄木鸟（1雄性、2雌性） 　 学名：Colaptes auratus 　 英文名：Golden-winged Woodpecker

PL. 255

Belted Kingfisher

Alcedo Alcyon

1 Males, 2 Female

Drawn from Nature by J.J.Audubon, F.R.S.F.L. Lith & Col.d by Bowen & Co. Philad.a

佩带翠鸟（1雄性、2雌性）

学名：Ceryle alcyon　英文名：Belted Kingfisher

PL. 275.

Yellow-billed Cuckoo

1 Male, 2 Female.

Papaw Tree

Drawn from Nature by J.J.Audubon, F.R.S.F.L.S. Lith.d Printed & Col.d by J.T.Bowen Philad.a

黄喙杜鹃（1雄性、2雌性）

学名：Coccyzus americanus　英文名：Yellow-billed Cuckoo

Mangrove Cuckoo.

Male.

Seven years apple.

Drawn from Nature by J.J.Audubon, F.R.S.F.L.S.　　Lith.d Printed & Col.d by J. T. Bowen, Philad.a

红树林杜鹃（雄性）　　学名：Coccyzus minor　英文名：Mangrove Cuckoo

Pl. 278.

Carolina Parrot or Parrakeet

1. 2. Males 3. Female. 4 Young.

Cockle bur.

Drawn from Nature by J. J. Audubon, F.R.S.F.L.S.

Lith.d Printed & Col.d by J. T. Bowen, Philad.a

马尾鹦鹉（1、2雄性、3雌性、4幼鸟）

学名：Conuropsis carolinensis　英文名：Carolina Parrot or Parrakeet

Band-tailed Dove or Pigeon?

1. Male. 2. Female.

Cornus Nuttallii!!

Drawn from Nature by J.J.Audubon.F.R.S.F.L.S.　　Lith^d Printed & Col^d by J. T. Bowen, Philad^a

带尾鸽（1雄性、2雌性）　　学名：Columba fasciata　英文名：Band-tailed Dove or Pigeon

Pl. 496.

The Texan Turtle Dove

Male.

Drawn from Nature by J.J.Audubon.F.R.S.F.L.S.

Lith.d Printed & Col.d by J.T. Bowen. Philad.a

德克萨斯斑鸠（雄性）

学名：Zenaida asiatica 英文名：Texan Turtle-Dove

Pl. 281.

Zenaida Dove.

1. Male 2. Female

Anona.

Drawn from Nature by J.J. Audubon F.R.S F.L.S.   Lith^d Printed & Col^d by J.T. Bowen, Philad^a

塞奈达野鸽（1雄性、2雌性）   学名：Zenaida aurita　英文名：Zenaida Dove

Pl. 282.

Key-West Dove.

Drawn from Nature by J.J. Audubon F.R.S.F.L.S. 1. Male 2. Female Lith^d Printed & Col^d by J.T. Bowen Philad^a

西威斯特鸽（1雄性、2雌性） 学名：Geotrygon chrysia 英文名：Key West Dove

Pl. 284.

Blue-headed Ground Dove or Pigeon.

Drawn from Nature by J.J. Audubon F.R.S.F.L.S. 1. Male 2. Females Lith^d Printed & Col^d by J.T. Bowen Philad^a

蓝头鸽（1雄性、2雌性） 学名：Starnoenas cyanocephala 英文名：Blue-headed Ground Dove or Pigeon

Pl. 285.

Passenger Pigeon.

1. Male. 2. Female.

Drawn from Nature by J. J. Audubon, F.R.S.F.L.S.　　Lith. & col. Bowen & Co. Philada.

候鸽（1雄性、2雌性）　学名：Ectopistes migratorius　英文名：Passenger Pigeon

Carolina Turtle Dove.

1. Males. 2. Females.

Drawn from Nature by J. J. Audubon, F.R.S.F.L.S.

Lith & col. Bowen & Co. Philada.

哀鸠（1雄性、2雌性）

学名：Zenaida macroura 英文名：Carolina Turtle-Dove

Pl. 287.

野火鸡（雄性）

学名：Meleagris gallopavo 英文名：Wild Turkey (Male)

Pl.288.

Wild Turkey. Female & Young

Drawn from Nature by J.J.Audubon, F.R.S.F.L.S　　Bowen & Co. lith & col. Philada

野火鸡（雌性和幼鸡）　　学名：Meleagris gallopavo　英文名：Wild Turkey (Female & Young)

Pl. 291.

Plumed Partridge.

Drawn from Nature by J.J.Audubon, F.R.S.F.L.S.　　1. Male 2. Female　　Lith^d Printed & Col^d by J.T. Bowen, Philad^a

刀翎鹑（1雄性、2雌性）　　学名：Oreortyx pictus　英文名：Plumed Partridge

冠齿鹑（幼鸟）　学名：Colinus cristatus　英文名：Welcome Partridge

环羽松鸡（1、2雄性、3雌性）　学名：Bonasa umbellus　英文名：Ruffed Grouse

Canada Grouse.

1.2 Males 3 Females.

Drawn from Nature by J.J.Audubon F.R.S.F.L.S. 4 Trillium pictum 5 Streptopus distortus. Lith.d Printed & Col.d by J.T. Bowen Philad.a

加拿大松鸡（1、2雄性、3雌性） 学名：Dendragapus canadensis 英文名：Canada Grouse

Pinnated Grouse.

Drawn from Nature by J.J.Audubon F.R.S.F.L.S. 1. 2. Males. 3 Female. Lilium Superbum. Lith.d Printed & Col.d by J.T. Bowen Philad.a

草原松鸡（1、2雄性、3雌性） 学名：Tympanachus cupido 英文名：Pinnated Grouse

Pl. 299.

Willow Ptarmigan

Drawn from Nature by J.J.Audubon.F.R.S.F.L.S. 1. Male. 2. Female & Young. Lith.d Printed & Col.d by J.T. Bowen Philad.a

柳雷鸟（1雄性、2雌性、3幼鸟） 学名：Lagopus lagopus 英文名：Willow Ptarmigan

Pl. 301

Rock Ptarmigan

1. Male, in Winter. 2. Female, Summer Plumage. 3. Young in August.

Drawn from Nature by J.J.Audubon.F.R.S.F.L.S. Lith.d Printed & Col.d by J.T. Bowen Philad.a

岩雷鸟（1雄性冬羽、2雌性夏羽、3幼鸟秋羽） 学名：Lagopus mutus 英文名：Rock Ptarmigan

Pl. 303.

紫鹁鸪（雄性春羽）　　学名：Porphyrula martinica　英文名：Purple Gallinule

Pl. 304.

鹁鸪（雄性）　　学名：Thomas Bewick　英文名：Common Gallinule

Pl. 309.

Great Red-breasted Rail, or fresh water Marsh Hen.

Drawn from Nature by J.J. Audubon, F.R.S.F.L.S. *1. Male adult. 2. Young.* Lith.d Printed & Col.d by J. T. Bowen, Philad.a

淡水秧鸡（1雄性、2幼鸟） 学名：Rallus elegans 英文名：Great Red-breasted Rail or Freshwater Marsh Hen

Drawn from Nature by J.J. Audubon, F.R.S.F.L.S. *1. Male. 2. Female. 3. Young.* Lith.d Printed & Col.d by J. T. Bowen, Philad.a

黑脸田鸡（1雄性、2雌性、3幼鸟） 学名：Porzana carolinus 英文名：Sora Rail

Whooping Crane.

Male, adult.

Drawn from Nature by J.J.Audubon, F.R.S.F.L.S.

Lith.d Printed & Col.d by J. T. Bowen, Philad.a

鸣鹤（雄性）

学名：Grusamericana 英文名：Whooping Crane

Pl. 314.

Whooping Crane.

Drawn from Nature by J.J.Audubon, F.R.S.F.L.S. Young. Lith.d Printed & Col.d by J. T. Bowen, Philad.a

鸣鹤（幼鸟） 学名：Grusamericana 英文名：Whooping Crane

Pl. 316.

American Golden Plover

1 Summer Plumage 2 Winter 3 Variety in March

Drawn from Nature by J.J.Audubon.F.R.S.F.L.S.

Lith.d Printed & Col.d by J.T. Bowen Philad.a

黑胸行鸟（1夏羽、2冬羽、3三月羽色）

学名：Pluvialis dominica　英文名：American Golden Plover

Pl. 317.

Kildeer Plover

Drawn from Nature by J.J.Audubon.F.R.S.F.L.S.

1 Male 2 Female

Lith.d Printed & Col.d by J.T. Bowen Philad.a

基尔第行鸟（1雄性、2雌性）

学名：Charadrius vociferus　英文名：Kildeer Plover

Pl. 318.

Rocky Mountain Plover

Drawn from Nature by J.J.Audubon,F.R.S.F.L.S. *Female.* Lith.^d Printed & Col.^d by J. T. Bowen, Philad.^a

山岩行鸟（雌性） 学名：Charadrius montanus 英文名：Rocky Mountain Plover

Pl. 320.

American Ring Plover

Drawn from Nature by J.J.Audubon,F.R.S.F.L.S. *1. Adult Male. 2. Young in August.* Lith.^d Printed & Col.^d by J. T. Bowen, Philad.^a

半蹼行鸟（1雄性、2秋季幼鸟） 学名：Charadrius semipalmatus 英文名：American Ring Plover

Pl. 321

Piping Plover

1 Male 2 Female

Drawn from Nature by J.J.Audubon F.R.S.F.L.S.

Lith^d Printed & Col^d by J. T. Bowen Philad^a

笛音行鸟（1雄性、2雌性）

学名：Charadrius melodus　英文名：Piping Plover

Pl. 323

Purple Gallinule

Drawn from Nature by J.J.Audubon F.R.S.F.L.S.

Adult Male Spring Plumage

Lith^d Printed & Col^d by J. T. Bowen Philad^a

翻石鹬

学名：Calidris canutus　英文名：Turnstone

Pl. 325

American Oyster-Catcher

Drawn from Nature by J.J. Audubon F.R.S.F.L.S. Male Lith.d Printed & Col.d by J. T. Bowen Philad.a

美洲蛎鹬（雄性）

学名：Haematopus palliatus　英文名：American Oyster-Catcher

Pl. 326.

Townsend's Oyster-catcher.

Drawn from Nature by J.J. Audubon F.R.S.F.L.S. Female. Lith.d Printed & Col.d by J. T. Bowen Philad.a

美洲黑蛎鹬（雌性）

学名：Haematopus bachmani　英文名：Townsend's Oyster-Catcher

Pl. 327.

Bartramian Sandpiper

Drawn from Nature by J.J.Audubon, F.R.S.F.L.S. 1. Male 2. Female. Lith.d Printed & Col.d by J. T. Bowen, Philad.a

丘陵矶鹬（1雄性、2雌性）

学名：Bartramia longicauda 英文名：Bartramian Sandpiper

Pl. 328.

Red-breasted Sandpiper

Drawn from Nature by J.J.Audubon, F.R.S.F.L.S. 1. Summer Plumage 2. Winter. Lith.d Printed & Col.d by J. T. Bowen, Philad.a

红胸矶鹬（1夏羽、2冬羽）

学名：Calidris canutus 英文名：Red-breasted Sandpiper

Pl. 330.

Purple Sandpiper.

1. Summer. 2. Winter.

Drawn from Nature by J.J. Audubon, F.R.S.F.L.S.   Lith.d Printed & Col.d by J. T. Bowen, Philad.a

紫矶鹬（1夏羽、2冬羽）   学名：Calidris maritima　英文名：Purple Sandpiper

Pl. 332.

红背矶鹬（1夏羽、2冬羽）   学名：Calidris alpina　英文名：Red-backed Sandpiper

Pl. 334.

Long-legged Sandpiper.

Drawn from Nature by J.J.Audubon.F.R.S.F.L.S. Lith^d Printed & Col^d by J.T.Bowen. Phila^d

长腿矶鹬 学名：Calidris himantopus 英文名：Long-legged Sandpiper

Pl. 337

Little Sandpiper.

Drawn from Nature by J.J.Audubon.F.R.S.F.L.S. 1. Male Summer Plumage. 2. Female. Lith^d Printed & Col^d by J.T.Bowen. Phila^d

小矶鹬（1雄性夏羽、2雌性） 学名：Calidris minutilla 英文名：Little Sandpiper

Pl. 339.

Red Phalarope

1. Adult Male 2. Winter plumage.

Drawn from Nature by J.J.Audubon.F.R.S.F.L.S.

Lith.d Printed & Col.d by J. T. Bowen, Philad.a

红瓣足鹬（1雄性、2冬羽）

学名：Phalaropus fulicaria 英文名：Red Phalarope

Pl. 340

Hyperborean Phalarope

Drawn from Nature by J.J.Audubon.F.R.S.F.L.S.

1. Male. 2. Female. 3. Young in autumn.

Lith.d Printed & Col.d by J. T. Bowen, Philad.a

红领瓣足鹬（1雄性、2雌性、3幼鸟秋羽）

学名：Phalaropus lobatus 英文名：Hyperborean Phalarope

Pl. 345.

Tell-tale Godwit or Snipe.

1. Male. 2. Female.

View of East Florida

Drawn from Nature by J.J.Audubon F.R.S.F.L.S. Lith^d Printed & Col^d by J. T. Bowen, Philad^a

大黄足鹬（1雄性、2雌性） 学名：Tringa melanoleuca 英文名：Tell-tale Godwit or Snipe

Drawn from Nature by J.J.Audubon F.R.S.F.L.S. *1. Male, 2. Female, 3. Young in Autumn.* Lith^d Printed & Col^d by J. T. Bowen, Philad^a

小山鹬（1雄性、2雌性、3幼鸟秋羽） 学名：Scolopax minor 英文名：American Woodcock

Pl. 354.

Black Necked Stilt.

Drawn from Nature by J.J.Audubon, F.R.S.F.L.S. *Male* Lith.d Printed & Col.d by J. T. Bowen, Philad.a

黑颈长脚鹬（雄性） 学名：Himantopus mexicanus 英文名：Black Necked Stilt

Pl. 356

Hudsonian Curlew.

Male.

Drawn from Nature by J.J.Audubon, F.R.S.F.L.S. Lith.d Printed & Col.d by J. T. Bowen, Philad.a

中杓鹬（雄性） 学名：Numenius phaeopus 英文名：Hudsonian Curlew

Glossy Ibis.

Adult Male.

Drawn from Nature by J.J.Audubon.F.R.S.F.L.S.

Lith.d Printed & Col.d by J. T. Bowen, Philad.a

彩鹮（雄性）

学名：Plegadis falcinellus 英文名：Glossy Ibis

Pl. 359.

Scarlet Ibis.

1. Adult male 2. Young second Autumn

Drawn from Nature by J.J.Audubon.F.R.S.F.L.S.

Lith.d Printed & Col.d by J. T. Bowen, Philad.a

红鹮（1雄性、2幼鸟秋羽）

学名：Eudocimus ruber 英文名：Scarlet Ibis

Wood Ibis

Male

Drawn from Nature by J.J.Audubon, F.R.S.F.L.S.

Lith.d Printed & Col.d by J. T. Bowen, Philad.a

树林鹳（雄性）

学名：Mycteria americana 英文名：Wood Ibis

Pl. 362.

Roseate Spoonbill

Male.

Drawn from Nature by J.J.Audubon.F.R.S.F.L.S. Lith.d Printed & Col.d by J. T. Bowen, Philad.a

玫瑰篦鹭（雄性）

学名：Ajaia ajaja 英文名：Roseate Spoonbill

Pl. 366

Great White Heron.

Male adult. Spring Plumage.

Drawn from Nature by J.J.Audubon.F.R.S.F.L.S. Lith.d Printed & Col.d by J. T. Bowen, Philad.a

大白苍鹭（雄性春羽）

学名：Ardea alba 英文名：Great White Heron

Pl. 364

Drawn from Nature by J.J.Audubon,F.R.S.F.L.S.　　Lith.d Printed & Col.d by J. T. Bowen, Philad.a

黄顶夜鹭（1雄性春羽、2 10月的幼鸟）　　学名：Nyctanassa violacea　英文名：Yellow-Crowned Night Heron

Least Bittern.

Drawn from Nature by J.J.Audubon,F.R.S.F.L.S. 1. Male 2.Female 3.Young. Lith.d Printed & Col.d by J. T. Bowen, Philad.a

小麻鸦（1雄性、2雌性、3幼鸟）

学名：Ixobrychus exilis 英文名：Least Bittern

Green Heron.

Drawn from Nature by J.J.Audubon F.R.S.F.L.S. 1 Adult Male. 2 Young in Sept.r Lith.d Printed & Col.d by J. T. Bowen, Philad.a

绿蓑鹭（1雄性、2 9月的幼鸟）

学名：Butorides virescens 英文名：Green Heron

Great blue Heron.

Male.

Drawn from Nature by J.J.Audubon.F.R.S.F.L.S.

Lith.d Printed & Col.d by J. T. Bowen, Philad.a

大蓝鹭（雄性）

学名：Ardea herodias　英文名：Great Blue Heron

Great American White Egret.
Male. Spring plumage

Drawn from Nature by J.J.Audubon F.R.S.F.L.S | Lith. & Col.d by Bowen & Co. Philada.

美洲大白鹭（雄性春羽）　学名：Casmerodius albus　英文名：Great American White Egret

Reddish Egret.
1.Adult, full Spring Plumage. 2. Young in full Spring Plumage two Years old

Drawn from Nature by J.J.Audubon F.R.S.F.L.S. | Lith & Col. by Bowen & Co. Philad.a

粉白鹭（1春羽、2幼鸟春羽）　学名：Egretta rufescens　英文名：Reddish Egret

Snowy Heron.

Drawn from Nature by J. J. Audubon, F.R.S.F.L.S. *Male* Lith. & Col^{d}. by Bowen & Co, Philada.

雪鹭（雄性） 学名：Egretta thula 英文名：Snowy Heron

PL 375

American Flamingo

Drawn from Nature by J.J. Audubon F.R.S.F.L.S　　Adult Male　　Lith & Col Bowen & Co Philada

美洲火烈鸟（雄性）　学名：Phoenicopterus ruber　英文名：American Flamingo

Canada Goose.

1. Male 2. Female.

Drawn from Nature by J.J.Audubon.F.R.S.F.L.S.

Lith^d Printed & Col^d by J. T. Bowen, Philad^a.

加拿大雁（1雄性、2雌性）

学名：Branta canadensis　英文名：Canada Goose

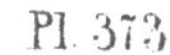

Louisiana Heron.

Male Adult.

Drawn from Nature by J.J.Audubo

Lithd Printed & Cold by J.T.Bowen, Philad.

三色苍鹭（雄性）

学名：Egretta tricolor　英文名：Louisiana Heron

Pl. 380.

White-fronted Goose.

1. Male. 2. Female.

Drawn from Nature by J.J.Audubon.F.R.S.F.L.S.

Lithd Printed & Cold by J.T.Bowen, Philad.

白额雁（1雄性、2雌性）

学名：Anser albifrons　英文名：White-fronted Goose

Pl. 382.

Trumpeter Swan

Adult.

Drawn from Nature by J.J.Audubon.F.R.S.F.L.S.

Lith.d Printed & Col.d by J. T. Bowen, Philad.a

号手天鹅（成鸟）

学名：Cygnus buccinator 英文名：Trumpeter Swan

Pl. 383.

Trumpeter Swan.

Young.

Drawn from Nature by J.J.Audubon.F.R.S.F.L.S.

Lith.d Printed & Col.d by J. T. Bowen, Philad.a

号手天鹅（幼鸟）

学名：Cygnus buccinator 英文名：Trumpeter Swan

Pl. 384.

American Swan

Male.

Drawn from Nature by J.J.Audubon.F.R.S.F.L.S. Lith.d Printed & Col.d by J. T. Bowen. Philad.a

小天鹅（雄性） 学名：Cygnus columbianus 英文名：American Swan

Pl. 385.

Mallarde

1. 2. Males. 3. 4. Females.

Drawn from Nature by J.J.Audubon.F.R.S.F.L.S. Lith.d Printed & Col.d by J. T. Bowen. Philad.a

绿头鸭（1、2雄性、3、4雌性） 学名：Anas platyrhynchos 英文名：Mallard

北美鸳鸯（1雄性、2雌性）

学名：Aix sponsa　英文名：Wood Duck-Summer Duck

Pl. 390

Pintail Duck.

Drawn from Nature by J.J. Audubon F.R.S.F.L.S. | 1. Male 2. Female. | Lith.d Printed & Col.d by J.T. Bowen, Philad.a

针尾鸭（1雄性、2雌性） 学名：Anus acuta 英文名：Pintail Duck

Pl. 393.

Blue-winged Teale.

Drawn from Nature by J.J. Audubon F.R.S.F.L.S. | 1. Male 2. Female. | Lith.d Printed & Col.d by J.T. Bowen, Philad.a

蓝翼短颈野鸭（1雄性、2雌性） 学名：Anas discors 英文名：Blue-winged Teal

Pl. 394

Shoveller Duck

1 Male 2 Female.

Drawn from Nature by J.J.Audubon F.R.S.F.L.S. Lith.d Printed & Col.d by J. T. Bowen Philad.a

琵嘴鸭（1雄性、2雌性） 学名：Anas clypeata 英文名：Shoveller Duck

Pl. 396

Red-headed Duck

1. Male. 2. Female.

Drawn from Nature by J.J.Audubon F.R.S.F.L.S. Lith.d Printed & Col.d by J. T. Bowen Philad.a

赤颈鸭（1雄性、2雌性） 学名：Aythya americana 英文名：Red-headed Duck

Pl 399.

Ruddy Duck.

1. Male. 2. Female. 3. Young.

Drawn from Nature by J.J.Audubon F.R.S.F.L.S.　　Lith.d Printed & Col.d by J. T. Bowen, Philad.a

棕尾硬鸭（1雄性、2雌性、3幼鸟）　　学名：Oxyura jamaicensis　英文名：Ruddy Duck

Pl. 405

Eider Duck.

1. Male. 2. Female.

Drawn from Nature by J.J.Audubon F.R.S.F.L.S.　　Lith.d Printed & Col.d by J. T. Bowen, Philad.a

绒鸭（1雄性、2雌性）　　学名：Somateria mollissima　英文名：Eider Duck

白斑头秋沙鸭（1雄性、2雌性）　学名：Mergellus albellus　英文名：White Merganser-Smew or White Nun

Pl. 411.

Buff-breasted Merganter Goosander.

Drawn from Nature by J.J.Audubon, F.R.S.F.L.S. 1. Male. 2. Female. Lith.d Printed & Col.d by J. T. Bowen, Philad.a

秋沙鸭（1雄性、2雌性） 学名：Mergus merganser 英文名：Buff-breasted Merganser or Goosander

Pl. 415.

Eider Duck.

1. Male. 2. Female.

Drawn from Nature by J.J.Audubon F.R.S.F.L.S. Lith.d Printed & Col.d by J. T. Bowen, Philad.a

普通鸬鹚（1雄性、2雌性、3幼鸟） 学名：Phalacrocorax carbo 英文名：Common Cormorant

Pl. 416.

Double-crested-Cormorant.

Male.

Drawn from Nature by J.J.Audubon.F.R.S.F.L.S. Lith.d Printed & Col.d by J.T.Bowen Philad.a

双顶鸬鹚（雄性） 学名：Phalacrocorax auritus 英文名：Double-crested Cormorant

Townsend's Cormorant

Male.

Drawn from Nature by J.J.Audubon,F.R.S.F.L.S

Lith.d Printed & Col.d by J.T. Bowen Philad.a

勃兰特鸬鹚（雄性）

学名：Phalacrocorax penicillatus 英文名：Townsend's Cormorant

Pl. 419.

Violet-green Cormorant.

Female in Winter.

Drawn from Nature by J.J.Audubon, F.R.S.F.L.S.

Lith.d Printed & Col.d by J. T. Bowen, Philad.a

海鸬鹚（雌性）

学名：Phalacrocorax pelagicus　英文名：Violet-green Cormorant

Pl. 420.

American Anhinga Snake Bird.

1. Male. 2. Female.

Drawn from Nature by J.J. Audubon, F.R.S.F.L.S. — Lith.d Printed & Col.d by J.T. Bowen, Philad.a

蛇鸟（1雄性、2雌性）　学名：Anhinga anhinga.　英文名：American Anhinga Snake Bird

Pl. 422.

Male.

Drawn from Nature by J.J.Audubon,F.R.S.F.L.S. Lith.d Printed & Col.d by J. T. Bowen, Philad.a

白鹈鹕（雄性） 学名：Pelecanus erythrorhynchos 英文名：American White Pelican

Pl. 426.

Booby Gannet.

Male.

Drawn from Nature by J.J.Audubon, F.R.S.F.L.S.

Lith.d Printed & Col.d by J. T. Bowen, Philad.a

褐鲣鸟（雄性）

学名：Sula leucogaster 英文名：Booby Gannet

Pl 424.

Brown Pelican

Young-first Winter.

Drawn from Nature by J.J.Audubon.F.R.S.F.L.S.

Lith.d Printed & Col.d by J.T.Bowen. Philad.a

褐鹈鹕（幼鸟）

学名：Pelecanus occidentalis 英文名：Brown Pelican

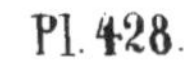

Black Skimmer or Shearwater.

Male.

Drawn from Nature by J.J.Audubon.F.R.S.F.L.S.

Lith.d Printed & Col.d by J.T.Bowen. Philad.a

黑剪嘴鸥（雄性）

学名：Rynchops niger 英文名：Black Skimmer or Shearwater

Pl. 430.

Gull-billed Tern - Marsh Tern.

Male.

Drawn from Nature by J.J.Audubon.F.R.S.F.L.S.

Lith.d Printed & Col.d by J. T. Bowen, Philad.a

鸥嘴燕鸥（雄性）

学名：Sterna nilotica　英文名：Gull-billed Tern or Marsh Tern

Pl. 453

Arctic Jager.

Drawn from Nature by J.J.Audubon.F.R.S.F.L.S. Lith.^{d} Printed & Col.^{d} by J. T. Bowen. Philad.^{a}

长尾贼鸥 学名：Stercorarius longicaudus 英文名：Arctic Jager

Common Tern.

Male Spring Plumage.

Drawn from Nature by J.J.Audubon, F.R.S.F.L.S.

Lith.d Printed & Col.d by J. T. Bowen, Philad.a

普通燕鸥（雄性春羽）

学名：Sterna hirundo 英文名：Common Tern

Roseate Tern.

Male

Drawn from Nature by J.J. Audubon, F.R.S.F.L.S.　　Lith & Col^d by Bowen & Co. Philad^a

红燕鸥（雄性）　　学名：Sterna dougallii　英文名：Roseate Tern

Pl. 440.

Noddy Tern.

Male.

Drawn from Nature by J.J.Audubon.F.R.S.F.L.S.

Lith.d Printed & Col.d by J. T.

玄燕鸥（雄性）

学名：Anous stolidus　英文名：Noddy Tern

Pl. 442.

Bonapartes Gull

1. Male in Spring. - 2. Female. - 3. Young first Autumn.

Drawn from Nature by J.J.Audubon.F.R.S.F.L.S.

Lith.d Printed & Col.d by J. T. Bowen. Philad.a

波拿巴鸥（1雄性春羽、2雌性、3幼鸟）

学名：Larus philadelphia　英文名：Bonaparte's Gull

Pl. 448.

Drawn from Nature by J.J.Audubon FRSFLS　　Lith.d Printed & Col.d by J.T. Bowen Philad.a

银鸥（1成鸟春羽、2幼鸟秋羽）　　学名：Larus argentatus　英文名：Herring or Silvery Gull

Great Black backed Gull.

Male.

Drawn from Nature by J.J.Audubon, F.R.S.F.L.S.

Lith.d Printed & Col.d by J. T. Bowen, Philad.a

大黑背鸥（雄性）

学名：Larus marinus 英文名：Great Black backed Gull

Glaucus Gull Burgomaster.

Adult male 2 Young first Autumn.

Drawn from Nature by J.J. Audubon, F.R.S.F.L.S.

Lith.d Printed & Col.d by J. T. Bowen, Philad.a

北极鸥（1雄性、2幼鸟）

学名：Larus hyperboreus　英文名：Glaucous Gull Burgomaster

Pomerine Jager

Adult Female.

Drawn from Nature by J.J. Audubon, F.R.S.F.L.S.

Lith.d Printed & Col.d by J. T. Bowen, Philad.a

中贼鸥（雌性）

学名：Stercorarius pomarinus　英文名：Pomerine Jager

Pl. 455.

Fulmar Petrel

Adult Male Summer Plumage.

Drawn from Nature by J.J.Audubon F.R.S.F.L.S. Lith.d Printed & Col.d by J.T. Bowen Philad.a

暴雪鹱（雄性夏羽） 学名：Fulmarus glacialis 英文名：Fulmar Petrel

Pl. 456.

Wandering Shearwater.

Male.

Drawn from Nature by J.J.Audubon F.R.S.F.L.S. Lith.d Printed & Col.d by J.T. Bowen Philad.a

灰鹱（雄性） 学名：Puffinus gravis 英文名：Wandering Shearwater

风头海鹦（1雄性、2雌性）

学名：Fratercula cirrhata 英文名：Tufted Puffin

普通海鹦鹉（1雄性、2雌性）

学名：Fratercula arctica 英文名：Common or Arctic Puffin

Razor-billed Auk

1 Male 2 Female

Drawn from Nature by J.J Audubon F.R.S.F.L.S

Lith.d Printed & Col.d by J. T. Bowen, Philad.a

刀嘴海雀（1雄性、2雌性）

学名：Alca torda 英文名：Razor-billed Auk

Pl. 467.

Curled-crested Phaleris.

Adult.

Drawn from Nature by J.J Audubon F.R.S.F.L.S.

Lith.d Printed & Col.d by J. T. Bowen, Philad.a

凤头海雀

学名：Aethia cristatella 英文名：Curled-crested Phaleris

Horned-billed Guillemot

Adult.

Drawn from Nature by J.J.Audubon,F.R.S.F.L.S. Lith.d Printed & Col.d by J. T. Bowen, Philad.a

角嘴海雀 学名：Cerorhinca monocerata 英文名：Horned-billed Guillemot

Pl. 476

Great North.n Diver Loon

1. Adult. 2. Young in Winter

Drawn from Nature by J.J.Audubon,F.R.S.F.L.S. Lith.d Printed & Col.d by J. T. Bowen, Philad.a

北潜鸟（1成鸟、2冬季幼鸟） 学名：Gavia immer 英文名：Great North Diver Loon

Pl. 478.

Red-throated Diver.

1. Male Summer Plumage 2, do Winter 3 Female. 4, Young.

Drawn from Nature by J.J.Audubon.F.R.S.F.L.S.

Lith^d Printed & Col^d by J. T. Bowen. Philad^a

红喉潜鸟（1雄性夏羽、2雄性冬羽、3雌性、4幼鸟）

学名：Gavia stellata 英文名：Red-throated Diver

Pl. 480

Red-necked Grebe.

1. Adult Male Spring Plumage 2. Young Winter Plumage

Drawn from Nature by J.J.Audubon.F.R.S.F.L.S.

Lith^d Printed & Col^d by J. T. Bowen. Philad^a

赤颈䴙䴘（1雄性春羽、2幼鸟冬羽）

学名：Podiceps grisegena 英文名：Red-necked Grebe

325幅鸟类图谱的说明

P13加利福尼亚美洲鹫

学名：Cathartes aura　英文名：Californian Turkey Vulture

加利福尼亚美洲鹫是北美洲一种普通的秃鹰，有着漂亮的红色面孔，灰色的羽毛，羽翼末端通常会有少量白色羽毛。它的飞行姿态非常优美，着陆时无须扇动翅膀。它有非常灵敏的嗅觉，是族群里唯一可以用嗅觉找寻动物尸体的秃鹰。加利福尼亚美洲鹫虽然是肉食鸟，但没有捕杀地面兽类的本领，而主要以地上的兽尸为食。

P14红头美洲鹫

学名：Cathartes aura　英文名：Red-headed Turkey Vulture

红头美洲鹫和加利弗尼亚美洲鹫属于同一种。它们有着相同的习性，不同的是红头美洲鹫的羽毛呈红褐色；雌性有橙红色的头，而雄性的头呈灰色。由于美洲鹫主要以地上的兽尸为食，随着工业的发达，狐兔之类的腐尸在大量减少，它的食物也相应减少；另外，它的繁殖力更弱，两年才生一个蛋，且成活率只有百分之五十。野生美洲鹫的现状可想而知。

P15上 黑秃鹰

学名：Coragyps atratus　英文名：Black Vulture or Carrion Crow

黑秃鹰的羽毛深暗，于是不可避免地会从外界吸收相当多的热量。它们的散热方式很特别：把尿撒在自己腿上，利用尿液蒸发使腿部降温，再依靠血液循环，为整个身体降温。

P15下 白头海雕

学名：Haliaeetus leucocephalus　英文名：White-headed Sea Eagle or Bald Eagle

白头海雕是一种大型猛禽。完全成熟的海雕体长可达1米，展开双翅可达2米多。它的视觉异常敏锐，即便翱翔于高空，也能洞察到地面、水中和树上的一切猎物。

P16长腿兀鹰

学名：Polyborus plancus　英文名：Caracara Eagle

长腿兀鹰有着红褐色的翅膀，尾羽有黄褐相间的条纹。和它的亲族一样，腐肉是长腿兀鹰们的最爱。长腿兀鹰属于猛禽类，和其他肉食猛禽一样，它有向下弯曲的钩形嘴，十分锐利，也有非常强健的足、锋利的爪。千万别小看了这些腐食动物的作用，它们就像清洁工一样及时清除腐败的动物尸体，从而有效地限制了疾病的传播和保证了水源的清洁。

P17栗翅鹰

学名：Parabuteo unicinctus　英文名：Harris's Buzzard

在鹰科这个大家族中，栗翅鹰的个头中等。火红色的翅膀和腿部的羽毛令栗翅鹰分外抢眼。栗翅鹰生活在热带和亚热带的半沙漠、半干旱林地和灌丛等干旱地区；常从栖木上突然飞起，猎食沙漠中的兔子、鼠、蛇、蜥蜴和鸟类。具有较明显的群体合作行为，与其他鹫鹰不同，它们常三只鸟共同捕捉和分享猎物，大家族群也会群体合作。

P18鵟

学名：Buteo buteo　英文名：Common Buzzard

鵟也叫土豹，外形像老鹰，但它的尾部羽毛不分叉，全身褐色，尾部颜色稍淡。常单独于空中翔鸣；栖息于平原、丘陵、海岸等开阔地及附近树林中，喜欢伫立于树梢。鵟属于肉食类猛禽，以昆虫、爬虫类、鼠类的活物为食。觅食时，鵟通常迎着风在空中定点振翅，一旦发现猎物，便迅速俯冲而下。

P19红尾鵟

学名：Buteo jamaicensis　英文名：Red-tailed Buzzard

红尾鵟属于鹰科鵟属，宽翅膀，宽尾巴。在美洲、欧洲和非洲可见到，中国台湾省也有发现。它们喜欢吃田地里的老鼠。红尾鵟在北美最为普遍，体长约60厘米，色彩较多。一般是上体呈褐色，下体颜色稍淡，尾巴呈红褐色。主要食物为啮齿动物，有时也吃小的哺乳动物、各种鸟类、爬行动物(包括响尾蛇、铜斑蛇)、两栖动物，甚至昆虫。

P20哈兰鹰

学名：Buteo Harlani　英文名：Harlan's Buzzard

哈兰鹰体型粗壮，双翅宽而钝圆，尾部宽阔，生活在农田、山丘、林间空地和森林边缘。哈兰鹰常在空中连续翱翔数小时搜寻猎物，或停栖在树上和电线杆上等候，一旦发现猎物立即猛扑而食，所捕捉的动物，小至甲虫、大至野兔，有时也吃小型腐尸。

P21红肩鹰

学名：Buteo lineatus　英文名：Red-shouldered Buzzard

红肩鹰是生活在开阔地带的典型猛禽，在上升的暖气流中盘旋翱翔搜寻猎物，也常停栖在较高的树上等候猎物，一遇到哺乳类、爬虫类、鸟类，及较大的昆虫即加以捕食。红肩鹰一般生活在农田和森林的混合地带以及半沙漠中，尤其是树木稀疏的地区。

P22宽翅鹰

学名：Buteo platypterus　英文名：Broad-winged Buzzard

宽翅鹰背部及双翼为褐色，胸部纵纹较稀疏，尾羽斑块密集，形成褐色斑带，胸腹部底色为淡白褐色。宽翅鹰翅膀宽广，可做持久滑翔飞行。翼末端黑色，当攻击小型猎物时，使猎物明显看到两边翼端闪动，逃走时不敢往两侧闪躲，以方便其从正中间攻击，提高攻击成功率。宽翅鹰还善于定点飞行，可在强风环境中猎食。

P23粗腿鹰

学名：Buteo lagopus　英文名：Rough-legged Buzzard

粗腿鹰以其粗壮的腿而得名。粗腿鹰的黄黑相间的花纹格外引人注目。它的喙呈钩状，异常锋利，视力极佳。它的腿粗壮有力，爪子锋利。粗腿鹰常常在草原湿地上低空翱翔寻找猎物。

P24金雕

学名：Aquila chrysaetos　英文名：Golden Eagle

金雕有黑色的虹膜，黄色的腊膜，灰色的喙，头颈上有金色的羽毛。黄色的脚上长满了羽毛，爪又大又强健，翼展长度达2.3米。金雕的窝筑在悬崖峭壁的洞穴里，或者孤零零的一棵大树上。金雕每窝大约产1-4个卵，一般是2个，雄雕和雌雕轮流孵化。金雕性情凶猛，经常袭击羊群，所以遭到牧人的捕杀。金雕现在数量稀少，处于濒危状态。

P25鱼鹰

学名：Pandion haliaetus　英文名：Common Osprey or Fish Hawk

鱼鹰属于冬候鸟。它头顶白色，腹部亦为白色，有黑褐色过眼带，翼型狭长，拍翅较缓，是以鱼为主食的猛禽。一般成熟的雄鸟约54厘米，雌鸟约64厘米。飞行时可见下身大部分为白色，这一特点使它很容易与其他鹰类区分。它经常栖息在湖泊、水库、池塘、河口与海湾，但迁移时可能通过山区。

P26上 游隼

学名：Falco peregrinus　英文名：Peregrine Falcon

游隼是一种猎鸟，体格强健，飞行速度很快。游隼在很高的空中飞行，一旦发现猎物，它会突然俯冲而下，抓起猎物用强健的脚掌狠击猎物的头部、背部，当猎物被击昏或击毙，从高空翻滚坠落时，游隼快速轻盈地跟着猎物下降，在半空中把猎物抓走。

P26下 燕尾鸢

学名：Elanoides forficatus　英文名：Swallow-tailed Hawk

燕尾鸢的尾羽分叉很大，形状和燕子酷似，因此而得名。它头比较小，脸上有点秃，短喙，双翼狭而长。燕尾鸢是一类体型较小的猛禽，善于飞行。

P27黑翅鸢

学名：Elanus caeruleus　英文名：Black-shouldered Elanus

黑翅鸢是小型猛禽，体长31～34厘米。喙为黑色，虹膜呈血红色。前额为白色，到头顶逐渐变为灰色。通常栖息于有树木和灌木的开阔原野、农田、树林和草原地区，喜欢单独活动，叫声细而尖。它在鼓翼飞翔时两翅扇动较轻，显得相当轻盈、优雅，然而当发现地面上的猎物时它会突然直扑而下。

P28密西西比鸢

学名：Ictinia mississippiensis　英文名：Mississippi Kite

密西西比鸢全身大致为灰色。头部、腹部、腿部为灰白色。翼下覆羽暗褐色。蜡膜及脚为黄色。虹膜为暗褐色。翼狭长后掠，适合于长时间盘旋觅食。尾长适中且内凹，可于飞行中灵活变换方向。翼下白斑可于远方吸引同类注意，或于攻击猎物时，使猎物注意两边而提高成功率。

P29矛隼

学名：Falco rusticolus　英文名：Iceland or Gyr Falcon

矛隼也叫白隼，是短距离飞行最快的鸟。隼科的猛禽都是些体形较小的鸟，它们中的大多数为候鸟。由于矛隼比鹰更易驯服，所以自古以来它就成了猎人的好帮手。另外一些有身份地位的人则用这种鸟来炫耀威武，也正是这种方式使得大量的矛隼惨遭厄运，已到了灭绝的边缘。冰岛有白色矛隼，数量极少，非常珍贵，是冰岛的国鸟。

P30灰背隼

学名：Falco columbarius　英文名：Merlin

灰背隼为小型猛禽，上体（包括两翼和尾等的表面）为蓝灰色，后颈有一道棕色领圈，杂以黑斑。喙短而勾，可增加咬合力。翼尖型，适于快速飞行。尾部稍长，适于空中急转弯，追击飞鸟。头部定位能力强，适于定点飞行时看清地上猎物。灰背隼形小性猛，飞行迅速，单独或成对活动。

P31雀鹰

学名：Accipiter nisus　英文名：Sparrow Falcon

雀鹰俗称鹞子，是鹰科中最小的猛禽，体长仅35厘米。主食鸟雀、鼠类，以善捕飞鸟而得名。当雀鹰发现猎物时，总是抢先飞到上方，占领制高点，然后收翅以每秒100米的速度，闪电般向猎物俯冲袭击，锐利的爪一下子抓住小鸟。更令人吃惊的是雀鹰竟能捕杀是它自身重量五倍的雉鸡。

P32苍鹰

学名：Accipiter gentilis　英文名：Gos Hawk

苍鹰属于中型猛禽，全长55厘米左右，身体细长，翅膀稍短，尾部呈分叉。它飞行速度缓慢，翅膀扇动的幅度大，能长时间滑翔，飞翔姿态优美。苍鹰栖息在森林之中，善于捕捉鼠、野兔等各种大小不同的动物，动作敏捷，灵活又凶猛。苍鹰分布地域广泛，长被训练用来打猎，它可以追逐和捕捉像狐狸和松鸡那样大的野生动物。

P33条纹鹰

学名：Accipiter striatus　英文名：Sharp-shinned Hawk

条纹鹰是一类个体小到中等的猛禽。它上体呈银灰色，下体呈白色，有棕色斑纹。尾比较长，有黑横纹和白色的尖端。它体长约30厘米，但以其勇猛和活跃著称。条纹鹰在高空和开阔地带飞行或迁徙时，飞翔速度很慢，显得缺乏力量，经过之处，有些飞得很快的欧洲八哥和燕科的鸟类会群集于它周围，佯装进攻。

P34灰泽鹫

学名：Circus cyaneus　英文名：Common Harrier

灰泽鹫雄鸟体上面及头部为灰色，翼末端为黑色。腹部、翼下面及尾下面白色。雌鸟全身以褐色为主，脸部有明显颜盘，头部及胸腹部均杂有密集纵纹，尾上有明显横纹，腰白色。灰泽鹫捕食时，双翼会展成V字形，在低空快速滑翔，搜索地面上的食物，这种姿势可惊起受伤的鸟、小型啮齿类和大型昆虫。

P35鹰鸮

学名：Ninox scutulata　英文名：Hawk Owl

鹰鸮体长约28～32厘米，头部呈圆形，体色大致为黑褐色。鹰鸮多在夜间活动，偶尔也在白昼活动，飞行能力很强，而且反应快速，会突击正在空中飞行的昆虫或其他动物，也猎食鸟类和小型哺乳类。它能生存于各种不同形态的森林，残存少数大树的开发地带，甚至海边的红树林都能发现它们的踪迹。

P36雪鸮

学名：Nyctea scandiaca　英文名：Snowy Owl

雪鸮别名白夜猫子，为大型猛禽，全长约60厘米。全身羽毛白色，有褐色斑点，头顶杂有少数黑褐色斑点。下体腹部具有窄的褐色横斑。嘴铅灰或黑褐色。爪灰褐色，末端黑色。雪鸮常栖息于冻土和苔原地带，也见于荒地丘陵。以鼠类、鸟类、昆虫为食。在北极和西伯利亚繁殖，越冬时可见于中国北方部分地区，十分罕见。

P37拉锯枭

学名：Aegolius acadicus　英文名：Saw Whet Owl

拉锯枭是密西西比河以东最小的猫头鹰，在美国多数地方和加拿大南部等地都可以见到。头棕色，眼部有白色条纹，体棕色，腹部有棕白相间的条纹。喜欢住在潮湿的地方和森林中。吃小哺乳动物和鸟类，大小和人的手掌差不多。拉锯枭是因其不一般的叫声得名的。据说它的叫声一般，如同单调的口哨声，但有时却像磨锯声。

P38仓鸮

学名：Tyto alba　英文名：Barn Owl

仓鸮的长相很特别，它虹膜周围的羽毛形成了心型的面盘。它能轻盈地自由滑翔，捕食时先在空地上方低飞或站在树枝上观察，然后无声地扑向猎物。在漆黑的夜间，它能藉声音来判断鼠类和各型小型哺乳动物的位置并加以捕捉。飞行时发出悠长颤抖的尖叫，筑巢时也会发出打鼾般的嘶嘶声。

P39灰鸮

学名：Strix nebulosa　英文名：Great Cinereous Owl

灰鸮的面盘很特别，淡褐色和黑色相间的条纹围绕虹膜一圈圈展开，形成苹果形状。背、双翼及尾羽均为褐色，杂有黑色花纹。腹部及腿部羽毛为淡褐色并杂有黑或深褐色的斑点。腿部粗壮有力，爪子锐利。灰鸮常在夜间活动，以小型哺乳动物和鸟类为食。

P40横斑林鸮

学名：Strix varia　英文名：Barred Owl

横斑林鸮有明显的褐色面盘，眼周的一条深褐色花纹酷似眉毛，鼻翼的深褐色花纹直达背部。喙和爪子均为黄色。全身褐色，背部有成横纹的淡褐色斑点，腿部有深褐色的竖纹。横斑林鸮平常多藏匿于草丛中，在阴天或黄昏时于低空盘旋，伺机捕食地上活动的小型哺乳类、鸟类和虫。

P41上 穴居鸮

学名：Speotyto cunicularia　英文名：Burrowing Owl

穴居鸮体长约24厘米，个头稍小。穴居鸮的面部为淡褐色，没有明显的面盘，全身为黑、白、褐色相杂的斑点。和其他猫头鹰不同的是穴居鸮擅长挖洞，喜欢穴居。

P41下 短耳鸮

学名：Asio flammeus　英文名：Short-eared Owl

短耳鸮体长之个体变异极大，约为33～43厘米左右。它是少数几种能在白天活动的猫头鹰之一，栖息于开旷的平原和泥泞的沼泽地。

P42巨角鸮

学名：Bubo virginianus　英文名：Great Horned-Owl

巨角鸮体形魁梧，体长约55厘米，是美洲体形最大的鸮类。它头部有耳状羽毛，腿部粗壮，爪子尖而长。巨角鸮通常以森林为家，而且会发出极低频的呼唤。这样低频的声音使得巨角鸮的呼唤可以在浓密的植被中传得较远。巨角鸮常利用喜鹊、乌鸦或其他猛禽的旧巢，孵化它的幼鸟，但有时也在树洞中筑巢。

P43小长耳鸮

学名：Strix asio　英文名：Little Screech-Owl

小长耳鸮是唯一有竖直耳状羽的中型褐色猫头鹰，体长约34厘米，具有明显的面盘，虹膜橙黄色，喙暗灰色。长耳鸮喜欢昼伏夜出，白天隐蔽在山地林间的树上或林中空旷的草丛中，夜幕降临时飞出觅食，飞行时毫无声息。小长耳鸮喜欢吃啮齿类动物，消化后毛、骨形成唾液吐出体外。图中的树种为松树。

P44夜鹰

学名：Caprimulgus　indicus　英文名：Night-Hawk

夜鹰属于夜行性攀禽。上体大都褐灰色，杂以黑褐色狭细横斑，头顶有黑色斑纹；胸灰白色，满布黑褐色横斑和虫斑；腹部、胁部均为红棕色，有浓密的黑褐横斑；尾下复羽棕白色，有淡褐色横斑。夜鹰白天喜欢伏在多树山坡的树杈处，夜间捕食昆虫。图中的树种为白橡树。

P45北美洲紫燕

学名：Progne subis　英文名：Purple Martin

北美洲紫燕体长约17厘米，喙张开时颇为宽大。背部及翼羽呈鲜艳之黑蓝色泽，腹部灰色。喜欢在树枝间筑巢，巢的形状像葫芦，较为封闭，悬挂在树枝上。有留鸟与过境鸟两种型态，常见于水塘、排水渠道、鸡鸭饲养场的上空。

P46树燕

学名：Tachycineta bicolor　英文名：White-bellied Swallow

树燕的头和背部呈蓝色，翅膀和尾羽为青灰色，腹部为白色。喙尖短，为黄色，爪子细小。在树燕群中，未生育的树燕们常在生育的成燕巢穴附近晃悠，想要伺机取而代之，这往往会演变成一场争夺巢穴的战斗。然而真正的主人经常会得到同伴的支援，侵略者就会知难而退。

P47岩燕

学名：Hirundo pyrrhonota　英文名：Cliff Swallow

岩燕的喉部为红褐色，额头黑色，喙的上部有一块白色羽毛，喉与腹部相交处有一白色环带。上胸有黑色横带，胸以下白色或黄白色。岩燕营巢于海岸峭壁的岩洞中或屋檐下，吞食海中小鱼、海藻或田野间的昆虫等小生物后，以唾液分泌物筑巢。这些巢穴经人工采集，选毛除污后便加工成了燕窝。

P48家燕

学名：Hirundo rustica　英文名：Barn or Chimney Swallow

家燕背部黑色有蓝色光泽，前额靠近嘴的部分呈暗橘红色。双翼长、尾羽有很深的分叉，常大角度来回飞行在空中，即能直接捕食蚊蚋与喝水。通常出现于平地至低海拔之空中或电线上。

P49紫绿燕

学名：Tachycineta thalassina　英文名：Violet-green Swallow

紫绿燕以身上丰富的色彩而得名。背及头部为绿色，背部的后半部分呈紫色，与前面的绿色截然分开。它的翅膀和尾羽呈黑褐色，羽毛有整齐的淡灰色边沿。腹部为白色。雄鸟的头顶为褐色。紫绿燕的虹膜为褐色，喙呈灰色，尖而短小。紫绿燕喜欢回旋飞翔，以小昆虫为食。

P50黄腹捕蝇鸟（雄性）

学名：Empidonax flaviventris　英文名：Yellow-bellied Flycatcher

黄腹捕蝇鸟的头部、背部和尾羽均为土灰色，它的虹膜呈褐色，有黄色的眼圈。它的喙为灰色，前额和咽喉有胡须状的羽毛。它的翅膀为黑色，有白色的横纹，翅尖的羽毛有白边。它的腹部呈黄褐色。

P51叉尾捕蝇鸟

学名：Tyrannus savana　英文名：Forked-tailed Flycatcher

叉尾捕蝇鸟身长约9厘米，它的显著特点是尾羽分叉而且极其细长，几乎是身体的3倍。它的头顶有一块褐色羽毛，背部为灰色，翅膀及尾羽为黑色，羽毛的边缘为褐色，将每根羽毛清晰地划分出来，腿部纤细，脚比较小。叉尾捕蝇鸟飞行时，尾巴上两根长长的外侧羽不断地左右摆动，一张一合，仿佛飞舞着一把活动的大剪刀。

P52剪尾王霸鹟

学名：Tyrannus forficatus　英文名：Swallow-tailed Flycatcher

剪尾王霸鹟有剪刀似的尾羽，是美国阿拉巴马州的州鸟。神气十足的抬头挺胸姿势是这种鸟的特征，它们以飞行在空中的小虫为食。鹟科鸟类往往有"定点捕食"的习性，剪尾王霸鹟也不例外，它们在一阵飞行捕食后会再飞回原本停驻的枝头，等待下一次的捕食机会。

P53阿肯色捕蝇鸟

学名：Tyrannus flaviventris　英文名：Arkansas Flycatcher

阿肯色捕蝇鸟头部有一块鲜红色的羽毛，背部、翅膀及尾羽为灰黑色，略微泛黄，尾羽的两端各有一根白色的羽毛，腹部呈黄色，腿爪纤小。阿肯色捕蝇鸟经常停在固定的枝头上，一见过往昆虫便飞过去拦截，再飞回原处，很少失手。这种捕食方式被称为"定点捕食"。

P54王鸟

学名：Tyrannus tyrannus　英文名：Tyrant Flycatcher or King-Bird

王鸟又叫必胜鸟，身长约20厘米。王鸟头部黑色，并有一块红色的羽毛，颈部深灰，腹部为白色，黑尾巴上镶有白色的边，黑白分明，别有一番风采。王鸟虽小，却十分勇敢，敢于和凶猛的庞然大物老鹰搏斗。图中的树木为白杨树。

P55凤头捕蝇鸟

学名：Myiarchus crinitus　英文名：Great Crested Flycatcher

凤头捕蝇鸟翼及尾羽呈褐色，背部是橄榄色，腹部为黄色，颈部呈淡蓝色。凤头捕蝇鸟喜欢居住在较开阔的环境，如大草原、农耕地、大型河川沿岸与市街区域，以昆虫为食。

P56菲比霸鹟（1雄性，2雌性）

学名：Sayornis saya　英文名：Say's Flycatcher

菲比霸鹟长相类似于捕蝇鸟，长期生活在水边，以小飞虫为食。它的喙为黑色，窄而尖。背部为褐色，翅膀黑色，羽毛的边缘有褐色或白色的窄条纹。尾羽黑色。腹部黄褐色，胸部呈苍白色。腿爪黑色。它们在悬崖的缝隙间筑巢。在遇到人之后总是显得很胆怯，而且比较沉默。

P57山岩捕蝇鸟（雄性）

学名：Sayornis nigricans　英文名：Rocky Mountain Flycatcher

山岩捕蝇鸟的头和背部就像黑烟煤熏黑了的褐色，翅膀和尾羽颜色稍深，二级羽和三级羽有白色的边。腹部白色，腿和爪子纤细锐利，为灰色。山岩捕蝇鸟和其他捕蝇鸟比起来似乎更善于利用翅膀，它能敏捷地捉住昆虫并咬断它们的翅膀。在袭击小飞虫时，它会使出各式各样的特技表演，就好像在空中舞蹈。图中的树木为槲栎。

P58阿卡迪亚捕蝇鸟（1雄性、2雌性）

学名：Empidonax virescens　英文名：Small Green-crested Flycatcher

阿卡迪亚捕蝇鸟的头部、背部为暗褐色，翅膀和尾羽的羽毛有淡褐色的边，咽喉灰白色，腹部呈微黄的白色。阿卡迪亚捕蝇鸟的喙为红色，眼周有黄色的眼圈。阿卡迪亚捕蝇鸟喜欢在一些干的小树枝上栖息，并耐心地观察着虹膜周围的物体，当它感觉到有一只昆虫就在附近的时候，就果断出击。图中的树木为檫木。

P59美洲小燕雀（1雄性、2雌性）

学名：Empidonax minimus　英文名：Pewee Flycatcher

美洲小燕雀的头顶有凸起的羽毛，酷似鱼鳍。背部和翅膀呈深褐色，翅膀的羽毛有淡褐色的边，腹部为白色。美洲小燕雀喜欢在树林中生活。与定点捕食的其他捕蝇鸟不同的是，有时美洲小燕雀在猎物与它有一定距离的时候会跟踪猎物，有时也会在树丛中寻找猎物。图中的植物为棉花。

P60跟踪捕蝇鸟（雄性）

学名：Empidonax traillii　英文名：Traill's Flycatcher

跟踪捕蝇鸟有不明显的顶饰，眼周有白色的眼圈。背部、翅膀及尾羽为黑褐色，翅膀的每级羽毛末端都呈蓝灰色。腹部为灰白色，腿爪为褐色，指甲尖锐。跟踪捕蝇鸟捕食的速度极快，能在瞬间捉住昆虫，然后又飞快地飞回刚才栖息的树枝上。图中的植物为橡胶树。

P61小燕捕蝇鸟（雄性）

学名：Empidonax minimus　英文名：Least Pewee Flycatcher

小燕捕蝇鸟眼周有白色的眼圈并与喙相连。背部、翅膀及尾羽为黑褐色，翅膀的羽毛边缘呈白色。腹部为黄色，脖子为灰白色，腿爪为灰色。主要栖息在森林边缘的橡树上。在休息的时候，它的羽冠会直立在头顶。小燕捕蝇鸟的叫声很洪亮，经常会持续鸣叫数小时。图中的植物为白橡树。

P62小头捕蝇鸟（雄性）

学名：Sylvania microcephala　英文名：Small-headed Flycatcher

小头捕蝇鸟的背部呈黄绿色，喙为黑色，眼周有白色的眼圈。它的翅膀和尾羽为黑色，并且很短，翅膀上有两条白色的横纹，腹部由黄色渐变为白色。小头捕蝇鸟喜欢在沼泽附近或水池边缘生活，在潮湿的气候下，它会显得异常活跃。图中的植物为曼佗罗花。

P63美洲红尾鸟（1雄性、2雌性）

学名：Tyrannus verticallis　英文名：American Redstart

美洲红尾鸟体长约11～14厘米。雄性头部、背部为蓝色，胸部和翅膀的中央呈橘色，尾羽为橘色，末端为黑色。雌性头部、背部呈黑色，咽喉白色。美洲红尾鸟十分活跃，它会在空中来回地旋转，橘色的羽毛在它的舞姿下更加迷人。图中的树木为铁栗木。

P64汤森宝石（雄性）

学名：Myadestes townsendi　英文名：Townsend's Ptilogonys

汤森宝石的头部、背部呈褐色，喙为黑色，眼周有灰蓝色的眼圈。它的翅膀和尾羽为黑色，并且很短，翅膀的羽毛边缘为灰白色。腹部呈褐色，到尾部渐变为黄褐色。汤森宝石擅长鸣叫，叫声清脆洪亮。

P65灰蓝色捕蝇鸟（1雄性、2雌性）

学名：Polioptila caerulea　英文名：Blue-grey Flycatcher

灰蓝色捕蝇鸟的头部、背部呈灰蓝色，喙为黑色。它的翅膀由灰蓝色渐变为黑色，并且很短，尾羽为黑色，向上翘起。腹部为白色，腿爪纤细。灰蓝色捕蝇鸟看起来小巧玲珑，颜色亮丽。它喜欢生活在森林里比较潮湿的地方或者沼泽地带。小溪、池塘或者河流的边缘也经常能发现它们的身影。图中的树木为黑胡桃树。

P66头巾林莺（1雄性、2雌性）

学名：Wilsonia citrina　英文名：Hooded Flycatching-Warbler

雄性头巾林莺的头部、腹部呈灰黄色，头顶及咽喉部分的黑色羽毛就像一块头巾系在头顶，喙为黑色。它的翅膀为绿色，并且羽毛上有黑边，尾羽为绿色，向上微翘，腿爪纤细。和雄性不同的是雌性没有黑色的头巾。头巾林莺总是在很低的小树杈上筑巢，苔藓、干草和草根是它们的建筑原材料。它们的巢穴精致而且紧密。图中的植物为虾脊兰。

P67加拿大捕蝇鸟（1雄性、2雌性）

学名：Wilsonia canadensis　英文名：Canada Flycatcher

雄性加拿大捕蝇鸟的头部、背部、翅膀及尾羽均呈褐色，喙为淡褐色。腹部黄色，有少许褐色斑点。雌性的背部颜色稍淡。加拿大捕蝇鸟喜欢居住在松树林中，飞行速度很快。当它在矮树丛中飞过时，你几乎看不到它。图中的植物为月桂树。

P68波拿巴林莺（雄性）

学名：Wilsonia canadensis　英文名：Bonaparte's Flycatching-Warbler

波拿巴林莺的头部、背部呈灰蓝色，喙为淡黄色，虹膜灰色，眼周有黄色的眼圈，喙角有胡须状的羽毛。它的翅膀的末端有两条白线，尾羽为灰蓝色。腹部为黄色，腿爪纤细。波拿巴林莺生性活跃，叫声清脆，在森林里经常能看到它活泼的身影。图中的植物为木兰花。

P69肯塔基林莺（1雄性、2雌性）

学名：Oporornis formosus　英文名：Kentucky Flycatching-Warbler

肯塔基林莺的头部、背部、翅膀及尾羽均为黄绿色，喙为黑色，虹膜黑色，眼周有黄色的眼圈，虹膜上部有一片柳叶形状的黄色羽毛。它的翅膀的羽毛有黑色边线。腹部为黄色，腿爪黄绿色。肯塔基林莺的巢很小，经常建在草茎上，巢建筑得很精美。图中的植物为玉兰花。

P70威尔逊林莺（1雄性、2雌性）

学名：Wilsonia pusilla　英文名：Wilson's Flycatching-Warbler

威尔逊林莺的头、背部呈灰绿色，喙为淡黄色，虹膜黄色，喙角有胡须状的羽毛，额头有一条黄色的羽毛一直延续到脖子。它的翅膀和尾羽为墨绿色，很短。腹部为黄色，腿爪为黄色。威尔逊林莺像其他捕蝇鸟一样以昆虫为食，善于定点捕食。它喜欢在湖边的树林栖息，在低矮的树丛中来回飞舞。图中的植物为蛇头草。

P71皇冠林莺（1雄性、2幼鸟）

学名：Dendroica coronata　英文名：Yellow-crowned Wood-Warbler

皇冠林莺的头、背部为褐色，喙为黑色，虹膜褐色。额头、尾部有一块黄色的羽毛。尾羽有白色斑点，向上翘起。雄性的身上多黑色斑纹，翅膀上有一块黄色的羽毛。雌性背部的斑点少，颜色深。皇冠林莺是一个飞行捕食的行家，除了飞虫以外，它还吃毛毛虫、草籽。在森林地带很少看到它的身影，它们喜欢在村舍附近的耕地活动。图中的植物为鸢尾花。

P72奥杜邦林莺（1雄性、2雌性）

学名：Dendroica coronata　英文名：Audubon's Wood-Warbler

奥杜邦林莺也是捕蝇鸟的一种，它的头呈灰蓝色，头顶有一块黄色的羽毛。背部呈深灰蓝色，有黑色斑点，喙为灰色，虹膜黄色，咽喉黄色。翅膀为灰蓝色，部分羽毛灰色。尾羽为黑色，有条白线。雄性腹部为黑色，渐变为白色，雌性腹部为淡灰色，腿爪纤细。奥杜邦林莺动作很优雅，常栖息于森林中。图中的植物为美洲卫矛。

P73黑顶白颊林莺（1雄性、2雌性）

学名：Dendroica striata　英文名：Black-poll Wood-Warbler

黑顶白颊林莺的头部为黑色、背部呈灰色有黑色斑点，喙为黑色，虹膜黑色，喙角有少量胡须状的羽毛。它的翅膀和尾羽为黑色，很短，羽毛上有白色的边线。腹部为白色，腿爪为淡黄色。黑顶白颊林莺的性情很温和，胆子大，不怕人。加拿大捕蝇鸟对它的威胁很大，它经常趁黑顶白颊林莺外出的时候去啄它的蛋。图中的植物为紫树。

P74海湾胸林莺（1雄性、2雌性）

学名：Dendroica castanea　英文名：Bay-Breasted Wood-Warbler

海湾胸林莺的头为黑色，头顶有一块褐色的羽毛。背部呈黑色，有褐色、白色及灰色的边线，喙为灰蓝色，虹膜褐色，咽喉褐色。尾羽为黑色，羽毛边缘有灰色边线。雄性脖子上有一块白色的羽毛，腹部为褐色，腹部中央为白色，雌性腹部为褐色，腿爪蓝灰色。海湾胸林莺喜欢吃棉铃虫，当棉花开满花朵，到处都能看到它忙碌的身影。图中的植物为丘陵棉花。

P75爬松林莺（1雄性、2雌性）

学名：Dendroica pinus　英文名：Pine Creeping Wood-Warbler

爬松林莺的头部、背部、喙为褐色，喙角有少量胡须状的羽毛。翅膀呈黑色，有白色的边线，虹膜褐色。尾羽为黑色，羽毛边缘有淡黄色边线。雄性的腹部为白色，雌性腹部为明黄色。爬松林莺的适应性很强，热带、温带、寒带的树林里都能发现它的身影，但它最喜欢栖息在松树林中。图中的植物为黄松。

P76芹叶钩吻林莺（1雄性、2雌性）

学名：Dendroica fusca　英文名：Hemlock Warbler

芹叶钩吻林莺也是捕蝇鸟的一种，它的头为黑色。背部呈深灰色，有黑色斑点，喙为明黄色，喙角黑色，虹膜黑色，咽喉为明黄色。它的翅膀为黑色，部分羽毛白色或有白边。尾羽呈黑色，有白色竖纹，腹部为明黄色渐变至白色，腿爪淡黄色。芹叶钩吻林莺只吃昆虫，喜欢栖息在茂密的树叶中。

P77黑喉绿林莺（1雄性、2雌性）

学名：Dendroica virens　英文名：Black-throated Green Wood-Warbler

黑喉绿林莺的头部及背部为黄黑相间的羽毛，眼周为明黄色，有黑色过眼线，虹膜为褐色。喙为黑色，有灰蓝色的边线。翅膀为黑色，羽毛有白色边纹。尾羽为黑色，羽毛边缘有白色边线。雄性的咽喉为黑色，腹部淡黄色。雌性的咽喉为黄黑相间的羽毛，腹部淡黄色。黑喉绿林莺以飞虫和毛虫为食。图中的植物为常青藤。

P78五月林莺（1雄性、2雌性）

学名：Dendroica tigrina　英文名：Cape May Wood-Warbler

五月林莺的头部为黑色，虹膜呈褐色，有黑色过眼线，喙为黑色。翅膀为黑色，羽毛有淡黄色边纹，部分羽毛白色。尾羽为黑色，并有白色宽横纹，羽毛边缘有黄色边线。雄性的眼部周围为红色。咽喉、腹部呈黄色，有黑斑。雌性的咽喉为黄黑相间的羽毛。五月林莺喜欢成双成对地活动。

P79黑斑林莺（1雄性、2雌性）

学名：Dendroica fusca　英文名：Blackburnian Wood-warbler

雄性黑斑林莺的头部、背部、喙呈黑色。翅膀呈黑色，有白色的边线，部分羽毛为白色，虹膜黑色，有黑色过眼线，眼周有黑色斑纹，咽喉、腹部为黄色，尾羽为黑色。雌性的头部、背部、喙为褐色，背部、眼周有深褐色斑纹，咽喉、腹部呈淡褐色。黑斑林莺的叫声婉转动听，以昆虫为食。图中的植物为草夹竹桃。

P80黄林莺（雄性）

学名：Dendroica petechia　英文名：Yellow-poll Wood-Warbler

黄林莺的头部为黄色，虹膜呈褐色，喙为黑色。背部、翅膀及尾羽为黄黑相间的羽毛。它的咽喉、腹部为黄色，有灰色斑点，尾羽为黄色，有黑色条纹，腿爪为淡黄色。黄林莺的叫声美妙动听，它总是唧唧喳喳地从一棵树跳到另一棵树上去捉小昆虫。它们喜欢把巢建在河流附近。

P81棕榈林莺（1雄性、2幼鸟）

学名：Dendroica palmarum　英文名：Yellow Red-poll Wood-warbler

雄性棕榈林莺的头部、虹膜为褐色，眼周呈黄色，有褐色或黄色过眼线。喙为褐色，喙角有少量胡须状羽毛。翅膀、背部及尾羽为灰色。咽喉、腹部为黄色。雌性的头部、喙为灰色。棕榈林莺偏爱橘子树的果园和天然的森林。最特别的是，棕榈林莺喜欢展开自己的尾羽，并不停地颤动。图中的植物为野橘树。

P82黄背蓝林莺（1雄性、2雌性）

学名：Parula americana　英文名：Blue Yellow-backed Wood-warbler

黄背蓝林莺的头部为蓝色，虹膜为黑色。喙呈淡黄色，喙角有少量胡须状羽毛。咽喉、背部为黄色，腹部的前面为黄色，后面呈白色。翅膀及尾羽为蓝色，有少量白色羽毛。黄背蓝林莺的羽毛色彩鲜艳，十分漂亮。它喜欢在池塘、湖泊或小河附近捕捉飞虫。花草的茎和秆是它的栖息之所。图中的植物为路易斯安那菖蒲。

P83汤森林莺（雄性）

学名：Dendroica townsendi　英文名：Townsend's Wood-Warbler

汤森林莺的头部为黑色，虹膜呈褐色，有黑色过眼线，眼部周围为黄色，喙为黑色。翅膀为黑色，羽毛有淡黄色边纹，部分羽毛白色。尾羽为黑色，并有白色宽横纹。咽喉为黑色，腹部为黄色，有黑斑，腹部中央为白色，腿爪黄色。汤森林莺经常在沼泽或小河边的树上栖息，喜欢停在树梢。图中的植物为卡罗莱纳多香果。

P84隐士林莺（1雄性、2雌性）

学名：Dendroica occidentalis　英文名：Hermit Wood-warbler

隐士林莺也是捕蝇鸟的一种，它的头为灰褐色，有黑色斑点。背部和咽喉呈深灰色，有黑色斑点，喙为黑色，虹膜褐色，有白色眼圈，脸颊为明黄色。它的翅膀为灰色，部分羽毛白色或有白边。尾羽为灰色，腹部为白色。隐士林莺十分胆小，容易受惊，喜欢单独活动。它喜欢隐藏在常绿树木的树叶深处。当它鸣叫的时候，通常有着固定的间隔。

P85黑喉灰林莺（雄性）

学名：Dendroica nigrescens　英文名：Black-throated Grey Wood-warbler

黑喉灰林莺的头部为黑色，虹膜呈褐色，有黑色过眼线，虹膜前方有一条明黄色的线，喙为灰色，脸颊为白色。背部呈灰蓝色，有黑色斑纹，翅膀为灰蓝色，有两条白色横纹。尾羽为黑色，并有白色宽竖纹。咽喉呈黑色，腹部为灰色，有黑斑，腹部中央为白色。黑喉灰林莺喜欢在橡树上筑巢，以橡树上的虫子为食。

P86黑喉蓝林莺（1雄性、2雌性）

学名：Dendroica caerulescens　英文名：Black-throated Blue Wood-warbler

雄性黑喉蓝林莺的头部、背部为灰蓝色，虹膜为深褐色，喙为黑色，脸颊、咽喉为黑色。翅膀为深灰蓝色，有白色斑纹。尾羽为黑色，并有蓝色和白色斑纹。腹部为白色。雌性黑喉蓝林莺的头部、背部、翅膀为墨绿色，虹膜上边有一条淡橙色的花纹，腹部为明黄色。黑喉蓝林莺是捕虫高手。

P87黑黄林莺（1雄性、2雌性、3幼鸟）

学名：Dendroica magnolia　英文名：Black & Yellow Wood-warbler

黑黄林莺的头部呈灰蓝色，虹膜为深褐色，有黑色过眼线，喙为黑色，咽喉为明黄色。它的背部呈黑色，翅膀为黑色，有白色斑纹，羽毛有白边。尾羽为黑色，并有白色斑纹。腹部呈黄色，有褐色斑点。黑黄林莺的巢经常筑在枞树的水平树枝上，以苔藓、草根、羽毛为原料，编织得很精细。

P88康涅狄格林莺（1雄性、2雌性）

学名：Oporornis agilis　英文名：Connecticut Warbler

雄性康涅狄格林莺的头部呈灰蓝色，背部为灰色，虹膜为黑色，喙为淡黄色，脸颊、咽喉呈黑色。它的翅膀为深灰色，尾羽为黑色，很短，腹部呈明黄色。雌性的头部、背部、翅膀为墨绿色，腹部为淡褐色，接近尾部的地方为明黄色。康涅狄格林莺在美国的康涅狄格州很常见，它们喜欢在地面上散步。图中的植物为皂草。

P89悲鸣林莺（1雄性、2雌性）

学名：Oporornis tolmiei　英文名：Mourning Ground-warbler

悲鸣林莺的头部为灰蓝色，虹膜呈褐色，喙为淡黄色，咽喉为黑蓝色渐变为黑色，蓝黑相交处有鱼鳞状花纹。它的翅膀、背部及尾羽呈黑黄色，腹部为明黄色，腿爪淡黄色。悲鸣林莺喜欢在低矮茂密的灌木丛栖息，偶尔也到地面上活动。图中的植物为雉眼草。

P90苦马沼泽林莺（雄性）

学名：Limnothlypis swainsonii　英文名：Swainson's Swamp Warbler

苦马沼泽林莺的头部、背部为褐色，虹膜呈淡褐色，喙为淡黄色，脸颊、咽喉为白色。它的翅膀为黑色，羽毛的边缘为褐色，尾羽为黑色，很短。它的腹部为白色，中央为黄色。腿爪为淡黄色，爪子纤细，修长。苦马沼泽林莺生活在沼泽地带，以水生植物上的虫子为食。它们经常栖息在水域附近的灌木丛中。图中的植物为金杜鹃。

P91食虫沼泽林莺（1雄性、2雌性）

学名：Helmitheros vermivorus　英文名：Worm-eating Swamp Warbler

食虫沼泽林莺的头部、咽喉及腹部为明黄色，虹膜呈褐色，有黑色过眼线，喙为黑色。它的头顶有两条黑色的竖纹，直通背部。背部、翅膀和尾羽为墨绿色，腿爪淡黄色。食虫沼泽林莺的鸟巢很特别，通常有两层。外层由干苔藓和山胡桃树或栗子树的树枝交错而成，内层是编织得很好的纤草。

P92书记湿地林莺（1雄性、2雌性）

学名：Protonotaria citrea　英文名：Prothonotary Swamp-warbler

书记湿地林莺的头为漂亮的黄色，喙为黑色，虹膜呈褐色。腹部为黄色，至尾部渐变成白色。背部深黄，至尾部渐变成灰蓝色。翅膀呈灰蓝色，尾羽黑白相间。腿爪为灰蓝色。书记湿地林莺经常在小溪及咸水湖附近的树林里栖息。它飞行的速度很快，当它飞过时，鲜艳的羽毛极其醒目。图中的植物为藤蔓。

P93巴赫湿地林莺（1雄性、2雌性）

学名：Vermivora bachmanii　英文名：Bachman's Swamp Warbler

巴赫湿地林莺的头、背部呈深褐色，脸颊、咽喉为黄色，虹膜褐色，喙黑色。它的翅膀呈黑色，羽毛边缘为褐色。尾羽呈黑色，羽毛边缘为灰蓝色。腿爪为褐色。雄性巴赫湿地林莺的胸部羽毛为黑色，有灰色横纹，雌性的胸部羽毛为褐色，有灰色横纹。巴赫湿地林莺很活跃，经常栖息在低矮的灌木丛中。图中的植物为茶花。

P94田纳西州湿地莺（雄性）

学名：Vermivora peregrina　英文名：Tennessee Swamp Warbler

田纳西州湿地莺的头部、背部呈淡褐色，虹膜呈褐色，有淡黄色的过眼线，喙为灰色。它的腹部为淡褐色，翅膀及尾羽为灰色，羽毛的边缘为淡褐色，尾羽短小。田纳西州湿地莺行动活跃敏捷，善于在飞行的时候捕捉昆虫。

P95蓝翅黄莺（1雄性、2雌性）

学名：Vermivora pinus　英文名：Blue-winged Yellow Swamp-warbler

蓝翅黄莺的头为黄色，虹膜呈褐色，有黑色过眼线，喙为黑色。它的背部为黄绿色，腹部为黄色。它的翅膀为褐色，初级羽和二级羽的末端为黄色，尾羽呈褐色，腿爪为黄色。蓝翅黄莺一般栖息在池塘附近的低矮树丛或草丛中，很活跃。它的巢穴由干苔藓和树叶构成，形状为少见的锥形。图中的植物为芙蓉葵。

P96黑白苔莺（雄性）

学名：Mniotilta varia　英文名：Black-and-white Creeping Warbler

黑白苔莺身上布满了黑色和白色的斑纹。它的动作优美迅速，歌声婉转动听。黑白苔莺寻找食物的时候非常有趣，它们经常在树干上蜷起一条腿做单脚跳。在保护它们的巢穴及幼鸟时，黑白苔莺有时会将一只翅膀张开，在地面可怜地呼叫并挣扎，就如同翅膀受伤了一般，以诱使掠食者远离它们的巢及小鸟。图中的植物为针叶松树。

P97褐色爬刺莺（1雄性、2雌性）

学名：Certhia americana　英文名：Brown Tree-Creeper

褐色爬刺莺的头、背部呈褐色，上面有整齐排成竖条的白色、淡褐色的斑点。它的面颊为淡褐色，虹膜为褐色，有褐色过眼线，喙尖长。它的翅膀呈灰褐色，并点缀着白色和淡褐色的横纹，腹部为淡褐色，尾羽的末端呈尖状。褐色爬刺莺栖息在森林里，它经常停留在高大的树木的树梢上。

P98岩石鹪鹩（雌性）

学名：Salpinctes obsoletus　英文名：Rock Wren

鹪鹩属雀形目，鹪鹩科。岩石鹪鹩体长约10厘米，体型小而矮胖，羽毛呈褐色，褐色的羽毛上有黑色的条纹。短喙稍微向下弯曲，翅膀短而圆，短尾巴向上竖起，腹部为淡褐色。鹪鹩在沼泽地、灌木丛和荒芜的山林中寻找昆虫来吃。它们飞到哪里就叫到哪里，歌声嘹亮。

P99卡罗莱纳州鹪鹩（1雄性、2雌性）

学名：Thryothorus ludovicianus　英文名：Great Carolina Wren

卡罗莱纳州鹪鹩的头和背部呈褐色，虹膜褐色，有淡褐色的过眼线。喙尖长，稍微向下弯曲。翅膀褐色，有黑色的斑纹，腹部淡褐色。尾羽的羽毛少。卡罗莱纳州鹪鹩属于短距离移栖动物，在美国卡罗莱纳州很常见。它们通常把窝安在灌木丛或突出的岩石上，雌鸟会在窝里铺上柔软的材料，使它们的家更舒适。

P100比威克鹪鹩（雄性）

学名：Thryomanes bewickii　英文名：Bewick's Wren

比威克鹪鹩的头部和背部为土黄色，虹膜呈褐色，有淡褐色的过眼线，喙为黑色。它的翅膀和尾羽有黑色的斑纹，腹部为淡灰色，腿爪灰褐色。比威克鹪鹩属于短距离候鸟，它经常把家安置在洞穴里，位置较低。以树上的昆虫为食。比威克鹪鹩的叫声嘹亮动听。

P101木鹪鹩（雄性）

学名：Troglodytes aedon　英文名：Wood Wren

木鹪鹩除了腹部以外全身大致为暗土黄色，背部有黑褐色细斑纹，翅膀、背部有黑褐色粗斑纹，腹部呈淡灰色，尾巴下面有灰白色斑点。木鹪鹩处于警戒状态时，常发出短促似“加、加”的声音。它生活在高海拔的草丛或树林底层。通常单独活动，喜欢鸣唱，生性隐匿，不容易看到。图中的植物为杨梅。

P102家鹪鹩（1雄性、2雌性、3幼鸟）

学名：Troglodytes aedon　英文名：House Wren

家鹪鹩体长约12厘米。它的头部、背部、翅膀及尾羽为褐色，翅膀和尾羽上有深褐色斑纹，尾羽短小，向上翘起。它的虹膜呈褐色，喙为深褐色，腹部为淡褐色。家鹪鹩属于热带地区的候鸟，以昆虫为食，在热带地区是比较普通比较常见的鸟。

P103冬鹪鹩（1雄性、2雌性、3幼鸟）

学名：Troglodytes troglodytes　英文名：Winter Wren

冬鹪鹩的头部和背部为深褐色，上面有黑色的细斑纹。它的面颊呈淡褐色，布满了深褐色的细斑纹，虹膜褐色，有不清晰的过眼线，喙为黑色。翅膀上有两条白色的横纹，尾羽短小，上翘。冬鹪鹩的腹部为淡褐色，有深褐色的细斑纹，腿爪为淡褐色。冬鹪鹩为短距离候鸟，生活在森林里，以昆虫为食。

P104泽地鹪鹩（1雄性、2雌性）

学名：Cistothorus palustris　英文名：Marsh Wren

泽地鹪鹩的头部为土黄色，上面有淡褐色的细斑纹。它的面颊呈淡褐色，有黑色的细斑纹，虹膜褐色，喙为黑色。尾羽上有黑色的斑纹，尾羽短小，上翘。它的腹部为淡褐色，有深褐色的细斑纹，腿爪呈淡褐色。雄性的头部有白黑相间的竖条纹，雌性的背部为黑色，有白色的花纹。泽地鹪鹩栖息在热带和温带的沼泽地带。

P105莎草鹪鹩（1雄性、2雌性）

学名：Cistothorus platensis　英文名：Short-billed Marsh Wren

莎草鹪鹩的头和背部为深褐色，有黑色斑纹，虹膜深褐色，有淡褐色的过眼线。翅膀褐色，有黑色的斑纹，腹部淡褐色，尾羽向上翘。腿爪纤细，为淡褐色。莎草鹪鹩生活在沼泽地的芦苇丛里，生性胆小。它的巢通常建在草丛上，用纤细的蓑衣草编织而成。

P106冠顶山雀（1雄性、2雌性）

学名：Parus atricristatus　英文名：Crested Titmouse

冠顶山雀的头部和背部为黑色，它与众不同的地方就是头顶有冠毛。它的脸颊呈淡灰色，虹膜褐色，喙为灰色。它的翅膀呈黑色，末端有淡灰色的羽毛，腹部、尾羽为灰色。冠顶山雀生活在森林里，经常栖息在比较隐蔽的树丛中。主要以各种昆虫以及幼虫为食，很少吃植物种子。图中的植物为白松。

P107黑顶山雀（1雄性、2雌性）

学名：Parus atricapillus　英文名：Black Cap Titmouse

黑顶山雀体长约12厘米。它的头部和咽喉为黑色，面颊呈白色，虹膜为黑色，喙为蓝灰色。它的背部为墨绿色，翅膀为黑色，羽毛的边缘为灰蓝色，尾羽呈灰蓝色，上翘。它的腹部呈淡褐色，中央为白色，腿爪为灰蓝色。黑顶山雀主要栖居在山脚溪河及平原河流两旁的松树林、松树和阔叶树混交林间以及城镇公园和风景区。图中的植物为山查树。

P108卡罗莱纳州山雀（1雄性、2雌性）

学名：Parus carolinensis　英文名：Carolina Titmouse

卡罗莱纳州山雀的头部呈黑色，虹膜为深褐色，喙为黑色，背部为褐色。它的翅膀及尾羽呈深灰色，腹部为淡褐色。雄性的面颊呈淡褐色，雌性为淡灰蓝色。卡罗莱纳州山雀平时成对地活跃在树枝间，不停地啄食。秋冬时节往往成群结队地活动。喜欢吃的食物为昆虫和草籽。

P109北山雀（1雄性、2雌性、3幼鸟）

学名：Parus hudsonicus　英文名：Hudson's Bay Titmouse

北山雀的头部和咽喉呈深褐色，面颊为白色，虹膜为褐色，喙为黑色。它的背部呈深褐色，翅膀为灰蓝色，尾羽为深褐色，上翘。它的腹部为淡褐色，中央呈白色，腿爪为灰蓝色。北山雀的鸣叫细柔婉转、多变，仔细琢磨十分有趣。北山雀经常把巢建在树洞、石垣和墙壁缝隙中。

P110赤背山雀（1雄性、2雌性）

学名：Parus rufescens　英文名：Chestnut Backed Titmouse

赤背山雀的头为黑色，虹膜深褐色，喙为黑色。它的面颊呈白色，咽喉为黑色。它的背部呈红褐色，翅膀为灰色，羽毛边缘呈白色。它的腹部为白色，脚爪为灰蓝色。赤背山雀栖息在森林里，它生性活跃，喜欢在树枝上跳来跳去。它的叫声清脆响亮，十分悦耳。

P111栗冠山雀（1雄性、2雌性）

学名：Psaltriparus minimus　英文名：Chestnut-crowned Titmouse

栗冠山雀的头部和背部为黑褐色，虹膜呈褐色。喙为黑色，周围有胡须状的羽毛，面颊为白色。它的翅膀和尾羽呈深褐色，羽毛的边缘为白色，腹部呈白色，腿爪为黑色。栗冠山雀的胆子很大，不怕人，它们总是栖息在低矮的树丛里或是森林里的灌木丛。它们用干苔藓、地衣筑巢，巢穴就像一个袋子一样悬挂在树枝上。

P112居维叶戴菊鸟（雄性）

学名：Regulus cuvieri　英文名：Cuvier's Kinglet

居维叶戴菊鸟是体形极小的莺科鸟类。它的外形很漂亮，黑色的头部带着一抹红色的条纹，白色的过眼线逐渐与背部的墨绿色融为一体。黑色的喙的周围长着少量胡须般的羽毛。墨绿色的翅膀和尾羽上点缀着黄色和白色的条纹。居维叶戴菊鸟活动在针叶林及针阔叶混合林中，冬季则有迁徙的现象。图中的植物为阔叶月桂树。

P113火冠戴菊鸟（1雄性、2雌性）

学名：Regulus satrapa　英文名：American Golden-crested Kinglet

火冠戴菊鸟的背呈橄榄绿色，翅膀及尾羽为黑褐色，有黄色宽横纹，体下污白色，嘴黑色，脚黄褐色，脸颊白色，眼周围黑色。雄鸟羽冠火红色，雌鸟黄色，但经常掩盖而呈黑色。它对树种的选择极为严格，仅停栖及觅食于针叶树，即使仅有一棵针叶树，也不会放弃。图中的植物为水生旅人蕉。

P114红冠戴菊鸟（1雄性、2雌性）

学名：Regulus calendula　英文名：Ruby-crowned Kinglet

红冠戴菊鸟的头部和背部呈灰绿色，头顶有一块红色的羽毛。它的虹膜为褐色，喙为黑色。翅膀和尾羽呈黑色，翅膀的羽毛上有白色和黄色的边纹。腹部为灰色，腿爪为黄色。红冠戴菊鸟生性活泼，常不断地在枝叶间跳跃，啄食昆虫。鸣唱声为由轻而重的“嘶——嘶——”声，极易误认为虫鸣声。图中的植物为绵羊月桂树。

P115知更鸟（1雄性、2雌性）

学名：Sialia sialis　英文名：Common Blue Bird

知更鸟的头部、背部、翅膀及尾羽均为蓝色。它的虹膜呈褐色，喙为黑色，咽喉为褐色。它的腹部为灰色，腿爪褐色。知更鸟经常停留在果园里，昆虫和水果都是它们的食物。知更鸟为留鸟，当冬天来临时，它们会成群结队地挤在一起睡觉，这样可以降低体内热量的消耗。图中的植物为毛蕊花。

P116西蓝鸲（1雄性、2雌性）

学名：Sialia mexicana　英文名：Western Blue Bird

和普通蓝鸲不同的是雄性西蓝鸲的背部和腹部为橙色，腹部中央呈白色，雌性西蓝鸲的头部和背部为灰色，面颊为橙色。西蓝鸲以橡树或低矮的松树为其栖息地，它们精力充沛，擅长鸣叫，但十分胆小，容易受惊。西蓝鸲是一雄一雌双宿双飞型的鸟，对配偶十分忠诚。

P117蓝知更鸟（1雄性、2雌性）

学名：Sialia currucoides　英文名：Arctic Blue Bird

雄性蓝知更鸟的头部、背部、翅膀及尾羽均为蓝色，翅膀的羽毛边缘为白色。它的虹膜为褐色，喙为深蓝色。腹部呈淡蓝色，腿爪为蓝色。雌性蓝知更鸟的咽喉及前腹为褐色，腹部的后半部分为白色。和同类知更鸟相比，它们更胆小。蓝知更鸟是有领域习性的独居动物，它们各自据巢为王，对于外来的鸟通常加以排斥。

P118北嘲鸟（1雄性、2雌性、3响尾蛇）

学名：Mimus polyglottos　英文名：Common Mocking Bird

北嘲鸟的头部和背部为墨绿色，虹膜为褐色，面颊为褐色，喙为黑色。它的翅膀为黑色，有灰色的边线，部分羽毛为白色。它的尾羽为黑色，有白色的边纹，腹部为褐色。北嘲鸟喜欢把巢建在常绿青藤上，用藤上的花朵来装饰它们的家，所以它们的巢总是充满芳香。不幸的是，它们的巢经常遭到蛇的袭击。

P119山嘲鸟（雄性）

学名：Oreoscoptes montanus　英文名：Mountain Mocking Bird

山嘲鸟的头部和背部呈土灰色，头部有深灰色的斑点，虹膜呈黄色，喙为黑色。它的翅膀和尾羽为深土灰色，上有白色的边纹。它的腹部为淡褐色，有深褐色的斑点，腿爪呈淡黄色。和其他嘲鸟一样，它善于模仿各种鸟类的叫声。它们经常在草丛或者低矮的灌木丛中筑巢，巢用小树枝和干草茎粗糙地搭建而成。它的卵为翠绿色。

P120褐嘲鸟（1、2、3雄性、4雌性）

学名：Toxostoma rufum　英文名：Ferruginous Mocking Bird

褐嘲鸟的头部、背部、翅膀及尾羽均为黄褐色，翅膀上各有两条有黑边的白色横纹。它的虹膜为黄色，喙为黑色。它的腹部为淡褐色，有黑色的斑点，腿爪为褐色。褐嘲鸟的巢经常遭到蛇的袭击，但是受害者通常会受到其他同伴的支援。褐嘲鸟不擅长长距离飞行，只能在矮树丛中飞来飞去，因为它们的翅膀短小。

P121迁徙画眉（1雄性、2雌性和幼鸟）

学名：*Turdus migratorius*　英文名：*American Robin or Migratory Thrush*

迁徙画眉的头部和背部为蓝黑色，虹膜和喙为黄色。它们以浆果为食，在吃一些草莓类浆果时，深红多汁的浆果汁经常会把它们的胸前染得血红。当冬季来临，食物减少时，它们就在农庄附近逗留，来寻找食物。它们经常吃楝树的果实，因此晕厥而从树上掉下来，所以很容易地就被抓住。图中的植物为板栗。

P122杂色画眉（1雄性、2雌性）

学名：*Ixoreus naevius*　英文名：*Varied Thrush*

杂色画眉的头部、背部和尾羽为灰褐色，头部和背部有淡褐色条纹。它的虹膜为白色，喙为黑色。杂色画眉栖居在山丘灌木丛和村落附近或城郊的灌木丛、竹林或庭院中。喜欢单独生活，秋冬结集小群活动。性机敏胆怯、好隐匿。常立树梢枝杈间鸣啭，引颈高歌，音韵多变、委婉动听，还善于模仿其他的鸟鸣声、兽叫声和虫鸣。图中的植物为桑寄生。

P123画眉（1雄性、2雌性）

学名：*Hylocichla mustelina*　英文名：*Wood Thrush*

画眉体长约24厘米。上体橄榄褐色，眼圈白色；下体棕黄色，有黑色斑点。画眉雌雄叫声明显不同，但外形却很难区分。画眉食性杂，但主要吃虫子，冬季吃一定数量的植物种子及果实。它们一般把巢建在地面草丛中、茂密树林或小树上。巢呈杯状或碟状，由树叶、竹叶、草、卷须等构成，内铺以细草、松针、须根之类。

P124茶色画眉（雄性）

学名：*Catharus fuscescens*　英文名：*Tawny Thrush*

茶色画眉的羽毛为褐色，腹部为淡褐色，有深褐色的斑点。茶色画眉生活在山区森林、村落附近的灌木丛或者竹林里。以昆虫，浆果或植物种子为食。它们喜欢单独或成对活动。茶色画眉雄鸟善鸣叫，歌声婉转流畅，极富韵味，声似“如意、如意”。雌鸟叫声为单调的“唧、唧、唧、唧”。图中的植物为山茱萸。

P125隐居画眉（1雄性、2雌性）

学名：*Catharus guttatus*　英文名：*Hermit Thrush*

隐居画眉的羽毛颜色为灰墨绿色，腹部为灰色，有黑色的斑点。它的虹膜为黄色，喙为土黄色。隐居画眉性情机敏，喜欢隐藏起来。但它们有保卫自己活动领域的习性，捕鸟者常利用此习性用养驯的画眉放在野生画眉的活动区域，引诱它来打斗，然后用活套捕捉它们。

P126金冠画眉（1雄性、2雌性）

学名：*Seiurus aurocapillus*　英文名：*Golden Crowned Wagtail (Thrush)*

金冠画眉的头部、背部、翅膀及尾羽均为深褐色，它的头顶有一块火红色的羽毛，因此而得名“金冠”。它的虹膜为黄色，有白色的眼圈，喙为深褐色。它的腹部为淡褐色，有深褐色斑点。金冠画眉成群穿梭于枝丫间，或攀附、倒悬于细枝上啄食小虫。主要以寄生在树皮裂缝中的昆虫为食，所以常见其在树干上下攀爬。图中的植物为蜀羊泉。

P127水栖画眉（1雄性、2雌性）

学名：*Seiurus motacilla*　英文名：*Aquatic Wood-wagtail*

水栖画眉的头部、背部、尾羽、虹膜及喙均为褐色，腿爪为淡粉色。雄性有一条火红色的眉线，它的腹部为黄色，上面点缀着深褐色的斑点。雌性的眉线为白色，腹部为淡黄色。水栖画眉生活在沼泽地带。它的叫声响亮、动听。它们把巢建在树根或者腐朽的木头旁边，用树叶和干苔藓堆砌而成，十分简陋。图中的植物为芜箐甘蓝。

P128密苏里州野云雀（雄性）

学名：*Sturnella magna*　英文名：*Missouri Meadow Lark*

密苏里州野云雀为候鸟，是开阔地区，特别是草原地带的一种代表鸟。密苏里州野云雀的头部为褐色和黑色相杂的斑点。它的虹膜呈褐色，有淡橙色的过眼线，它的喙为黑色。它的背部、翅膀呈褐色，遍布黑色的斑点。它的尾羽为灰色，腹部呈黄色，有黑色的斑纹。密苏里州野云雀以野生植物种子、昆虫为食，巢多营于地面陷凹处，善歌咏。

P129上 海滨百灵（1雄性夏羽、2雄性冬羽、3 雌性、4幼鸟）

学名：*Eremophila alpestris*　英文名：*Shore Lark*

百灵鸟又叫云雀、告天子、朝天柱，它属雀形目，百灵科，体形和羽毛极像麻雀，是著名的鸣禽和笼养鸟。海滨百灵毛并不华丽，但雄鸟歌声音韵婉转，深受人们喜爱。

P129下 拉普兰白頬百灵（1雄性春羽、2雄性冬羽、3 雌性）

学名：*Calcarius lapponicus*　英文名：*Lapland Lark Bunting*

拉普兰白頬百灵的羽毛布满黑褐相间的斑点，虹膜为黄色。雄性百灵鸟会随着季节的变化改变头部羽毛的颜色。经过训练的百灵鸟能模仿各种动物的叫声。

P130灰雀（雄性）

学名：*Spizella pallida*　英文名：*Shattuck's Bunting*

灰雀的体长为14厘米左右，雌雄形、色非常接近。喙褐色，呈圆锥状；头、颈灰蓝色，背部为深褐色，饰以黑色条纹。脸頬呈淡褐色，左右各有褐色斑条，腹部也为淡褐色。尾羽为黑色。灰雀多活动在有人类居住的地方，性极活泼，胆大易近人，但警惕性却非常高，好奇心较强。多筑巢于树洞中。

P131尖尾鹀（雄性）

学名：*Passerherbulus cauda cutus*　英文名：*Le Conte's Sharp-tailed Bunting*

尖尾鹀体长约15厘米，属普遍的冬候鸟。它的喙粗短，呈灰蓝色。它的面頬及背部、翅膀的羽毛呈黄褐色，并杂有黑色的斑点，它的尾羽黑褐色，腹部白色。尖尾鹀栖息于平地至中海拔山区开垦地、农耕地、草丛地，通常成群甚至大群地出现在杂草、芒草中觅食禾本科植物种子。

P132白頬百灵（雄性）

学名：*Calcarius pictus*　英文名：*Painted-lark Bunting*

白頬百灵属于中小型鸣禽，体长17～19厘米。头与背部呈深褐色，间杂有褐色斑纹，眼四周有白色花纹，颈部腹部为黄褐色，喙为黑色，足为黄褐色。白頬百灵以草籽、蝗虫、老鼠等为食。营巢于草原或沙滩凹处。它善于鸣叫，鸣声嘹亮宽广，婉转多变。传说能学百声，故称“百灵”。

P133赤颈百灵（雄性）

学名：*Calcarius ornatus*　英文名：*Chestnut-collared Lark Bunting*

赤颈百灵的头部为黑色，面頬有白色斑纹，虹膜和喙均为褐色。它的颈部有一块红色的羽毛，咽喉为淡褐色。它的背部和翅膀为土黄色，有黑色的斑点，尾羽为黑色。它的腹部为黑色，至尾部为白色。赤颈百灵善于鸣叫，歌声嘹亮，而且它行动敏捷，飞翔很快，飞行时且飞且唱。

P134雪云雀（1、2成鸟、3幼鸟）

学名：*Plectrophenax nivalis*　英文名：*Snow Lark Bunting*

雪云雀的头部为褐色，有白色的花纹，它的面頬为白色，虹膜为红色，有褐色的过眼线，喙为淡黄色。它的背部为褐色，有黑色的斑纹，翅膀和尾羽为白色，部分为黑色。它的咽喉、腹部均为白色，腿爪为黑色。雪云雀生活在草原地带，食物以野生植物果实和昆虫为主，特别喜吃蝗虫幼虫。

P135汤森云雀（雄性）

学名：*Emberiza townsendi*　英文名：*Townsend's Bunting*

汤森云雀的头部为淡绿色，带有细小的黑色条纹，面頬和咽喉为白色，虹膜为黄色，有与头部相同颜色的过眼线，喙为淡蓝色。它的背部、翅膀以及尾羽为褐色，上面点缀着深褐色的斑纹，腹部为淡褐色，腿爪为淡蓝色。汤森云雀以食地面上的昆虫和种子为生。所有的云雀都有高昂悦耳的声音。

P136湾翼麻雀（雄性）

学名：*Pooecetes gramineus*　英文名：*Bay-winged Bunting*

湾翼麻雀的头为灰褐色，上面有黑色斑点，面頬白色。它的虹膜为褐色，喙短，为圆锥状，呈淡褐色。它的背部、翅膀以及尾羽均为黑褐色，翅膀的根部有一块橙色的羽毛，两条淡褐色的横条纹，腹部为灰白色，有少量黑色的斑点。湾翼麻雀以草籽及其他植物种子为食，兼食昆虫；常在地面跳跃行进，啄食地面食物。图中的植物为仙人掌。

P137草原麻雀（1雄性、2雌性）

学名：*Passerculus sandwichensis*　英文名：*Savannah Bunting*

草原麻雀的身上布满了黑色与淡黄色的花纹，它的面頬为淡黄色，虹膜为褐色，喙为深褐色，尾羽为黑色，很短。它的腹部为淡黄色，有少量黑色的斑点，腿爪为淡褐色。草原麻雀经常成群结队地栖息在平地、丘陵的小河、湖泊、池塘等水域旁的草丛。叫声为轻细的“啧、啧”声，声音短促。

P138土色麻雀（雄性）

学名：*Spizella pallida*　英文名：*Clay-colored Bunting*

土色麻雀的头部、背部及尾羽为土灰色，头部和背部布满了黑色的斑点，翅膀和尾羽的羽毛有淡土黄色的边。它的虹膜为土黄色，喙为淡黄色。它的腹部为淡黄色，腿爪为土黄色。土色麻雀就栖息在人类的活动区内，通常在民房的屋檐下或是通风口处筑巢；在秋、冬季时，常大群地聚集在芦苇林、草地或是稻田附近。图中的植物为马雷筋。

P139黄翼麻雀（雄性）

学名：*Ammodramus savannarum*　英文名：*Yellow-winged Bunting*

黄翼麻雀的头部和背部为灰蓝色，上面有黑色的斑点，它的虹膜和喙为褐色，眼睛上面有一条黄色的羽毛，酷似眉毛。它的翅膀为褐色，羽毛上有黄色、橙色和紫色的边，尾羽为黑色，有褐色的边，腹部为淡黄色。黄翼麻雀常成小群地在灌木丛、林缘与农耕地活动，被惊吓时，成群飞入灌木丛上层。它以植物种子及果实、昆虫为食。

P140亨斯麻雀（雄性）

学名：*Ammodramus henslowii*　英文名：*Henslow's Bunting*

亨斯麻雀体长约14厘米，头部和背部为暗土黄色，有黑色斑点，尾羽黑褐色，虹膜为褐色，喙为土黄色。它的腹部为淡黄色，胸前有黑色小斑点。亨斯麻雀喜群栖、爱喧哗，不惧怕人，喜停栖于屋顶、电线上或地面。亨斯麻雀脚短，以蹦跳方式在地面上移动。图中的植物为福禄花。

P141田间麻雀（雄性）

学名：*Spizella pusilla*　英文名：*Field Bunting*

田间麻雀头部和背部为土绿色，翅膀为墨绿色，上面有一条淡黄色的横条纹，部分羽毛的边缘为褐色，尾为淡灰绿色。它的虹膜为暗褐色，腹部为淡灰色。田间麻雀族群数量庞大，农村收割时节，常见形成上百只群栖于谷场或电线上，为平地数量最多，最普遍的鸟种。图中的植物为越橘和草粉兰。

P142碎屑麻雀（雄性）

学名：Spizella passerina　英文名：Chipping Bunting

碎屑麻雀的头部和背部为红褐色，上面有整齐的黑褐色斑点，它的虹膜和喙为红褐色，面颊为白色，有黑色的过眼线。它的翅膀为黑色，羽毛边缘为白色或者红褐色，尾羽为红褐色，腹部为白色。碎屑麻雀也是比较常见的一种小鸟，它们经常在屋子前面、树枝上，甚至道路上轻巧地蹦来跳去。图中的植物为刺槐。

P143灯心草雀（1雄性、2雌性）

学名：Junco hyemalis　英文名：Oregon Snow Bird

灯心草雀的头部、面颊以及咽喉为黑色，虹膜和喙为黄色，它的背部、翅膀和尾羽为暗褐色，翅膀上的部分羽毛的边缘为白色。它的腹部为白色，腿爪为淡黄色。灯心草雀在农户的屋檐下或墙壁裂缝里栖息。它主要以谷物、草籽和昆虫为食，也喜欢吃人类丢弃的各种食物。不过它们在喂幼雏时，一定要捕捉昆虫。图中的植物为金樱子。

P144五彩雀（1、2、3雄性、4雌性）

学名：Passerina ciris　英文名：Painted Bunting

五彩雀的羽毛色彩艳丽，雄性的头部为蓝色，咽喉和腹部为水红色，它的背部前面为草绿色，至后面颜色渐黄，背部的后面为水红色，翅膀为墨绿色和深褐色，尾羽深褐色。雌性的头部、背部、翅膀和尾羽为绿色，腹部为黄色。五彩雀为热带地区的候鸟，以昆虫和草籽为食。图中的植物为野李子。

P145青雀（1、2、3雄性、4雌性）

学名：Passerina cyanea　英文名：Indigo Bunting

雄性青雀在不同的地域，毛色稍有差异，大体来说一般为靛蓝色，在暗淡的光线下颜色有点发黑。翅膀的羽毛有少量灰色，有些颜色稍淡，灰色羽毛较多。雌性的头部和背部为黑褐色，面颊和腹部为褐色，翅膀和尾羽为黑色。青雀为热带地域的候鸟，以昆虫为食，也吃植物的种子和水果。图中的植物为野生撒尔沙。

P146天青石雀（1雄性、2雌性）

学名：Passerina amoena　英文名：Lazuli Finch

雄性天青石雀的头部、咽喉、背部和尾羽均为蓝色，虹膜为褐色，喙为淡蓝色。它的翅膀为黑色，羽毛的边缘为蓝色，翅膀上有两条白色的横条纹。它的腹部为黄褐色，至尾部渐变为白色。雌性的头部、背部为深褐色，腹部为褐色。天青石雀生性活泼，歌声动听，经常把巢安置在柳树上。图中的植物为野生西班牙咖啡树。

P147海滨雀（1雄性、2雌性）

学名：Ammodramus maritimus　英文名：Sea-side Finch

雄性海滨雀的头部、背部和腹部为灰色，上面点缀着细小的黑色短条纹。它的虹膜为褐色，眼睛上面有一条黄色的眉毛状羽毛。它的翅膀和尾羽为黄褐色，尾羽的羽毛末端为尖状。雌性的头顶为红褐色。海滨雀栖居在海滨或者咸水沼泽地，它们喜欢鸣叫，很少有安静的时候。它们以昆虫，小的螃蟹和蜗牛为食，也吃绿色沙甲虫。图中的植物为卡罗莱纳玫瑰。

P148尖尾雀（1雄性、2雌性）

学名：Ammodramus caudacutus　英文名：Sharp-tailed Finch

尖尾雀的头部为黑色，面颊为橙色，虹膜为褐色，有蓝灰色的过眼线，喙为灰色。它的背部为黑色，羽毛的边缘为白色，翅膀和尾羽为黑褐色，尾羽短小，整体形状呈尖状。它的腹部为淡灰色，有少量黑色的斑点。尖尾雀生活在湿地附近，它们把巢穴建在地面上，用蓑衣草编织而成。

P149湿地麻雀（雄性）

学名：Melospiza georgiana　英文名：Swamp Sparrow

湿地麻雀的头部为深褐色，面颊为淡褐色，虹膜为褐色，有褐色的过眼线，喙为深褐色。它的背部为淡褐色，上面有深褐色的斑点，翅膀和尾羽为深褐色，有淡褐色的斑点或者边纹。它的腹部为淡灰色，腿爪为褐色。湿地麻雀生活在淡水沼泽地，以甲虫、蟋蟀、蚱蜢为食。湿地麻雀的卵为蓝绿色，有褐色的大斑点。图中的植物为八角莲。

P150巴赫松雀（雄性）

学名：Aimophila aestivalis　英文名：Bachman's Pinewood Finch

巴赫松雀的头部和背部为黑褐色，上面有黑色的斑点，头顶有淡褐色的竖条纹，它的虹膜为黄褐色，有淡褐色的过眼线，它的喙为黑褐色。它的翅膀和尾羽为黑色，翅膀的羽毛边缘为灰色。它的腹部和腿爪为淡褐色。巴赫松雀习惯于站在高大的松树上鸣叫，歌声美妙动听。它以草籽、甲虫和浆果为食。

P151林肯松雀（1雄性、2雌性）

学名：Melospiza lincolnii　英文名：Lincoln's Pinewood-Finch

林肯松雀的头部、背部和尾羽为褐色，头部和背部有黑色的斑点。它的虹膜为红褐色，有淡蓝色的过眼线，它的喙为淡蓝色，面颊为淡褐色。林肯松雀生活在低矮的灌木丛中，它喜欢在树枝头鸣唱。它性格暴躁好争斗，两只雄雀相遇，通常要争斗半天直到一方疲劳逃走为止。图中的植物1是山茱萸，2是野生黄莓，3是蓝绿色石南。

P152粉末金翅鸟（雄性）

学名：Carduelis hornemanni　英文名：Mealy Redpoll Linnet

粉末金翅鸟的头顶有一块水红色的羽毛，它的面颊为淡粉色，虹膜为黄色，喙为黄色，咽喉为黑色。它的背部为黄褐色，有黑色的斑点。它的翅膀和尾羽为黑色，羽毛边缘为白色。它的腹部淡粉色，两翼附近有黑色的斑点。粉末金翅鸟生活在树林里，以浆果和草籽为食。

P153小红雀（1雄性、2雌性）

学名：Carduelis flammea　英文名：Lesser Redpoll Linnet

小红雀的头部、背部、翅膀和尾羽为土灰色，它的头顶有一块水红色的羽毛。它的虹膜为黑褐色，喙为黄色。雄鸟的腹部为水红色，有少量细小的黑色斑点，尾羽的跟部也是水红色。雌鸟的腹部为淡灰色，有黑色的斑点。小红雀栖息在矮树丛中，以浆果为食，在冬季，它们一般栖息在河流或湖泊岸边的芦苇丛中。

P154松红雀（1雄性、2雌性）

学名：Carduelis pinus　英文名：Pine Linnet

松红雀的头部和背部为土绿色，上面布满了黑色的斑点。它的虹膜为深褐色，喙为黑色。它的翅膀和尾羽为墨绿色，羽毛的边缘为淡黄色。它的腹部为淡黄色，遍布墨绿色的斑点。松红雀像一个流浪者一样在不同的山脉间以不规则的周期流浪，它们的性情大胆而温和，当站在树枝头的时候，它们喜欢把头低下，向下张望。图中的植物为黑落叶松。

P155美洲金翅鸟（1雄性、2雌性）

学名：Carduelis tristis　英文名：American Goldfinch

雄性美洲金翅鸟的头部为黑色，虹膜和喙为褐色。它的背部和腹部为明黄色，翅膀和尾羽为黑色，有少量白色羽毛。腿爪为褐色。雌鸟的喙、头部和背部均为深褐色，翅膀的部分羽毛为墨绿色。美洲金翅鸟为短距离迁徙鸟，它喜欢栖息在果园或者小树林里，以植物的种子为食。图中的植物为蓟。

P156黑头金翅鸟（雄性）

学名：Carduelis magellanica　英文名：Black-headed Goldfinch

黑头金翅鸟的头部和咽喉为黑色，虹膜为褐色，喙为淡蓝色。它的背部和腹部为黄色，背部中央为草绿色，上面有黑色的小细纹。它的翅膀和尾羽为黑色，翅膀的部分羽毛为黄色，或有黄色的边。它的腿爪为淡黄绿色。黑头金翅鸟的食物主要是树木和杂草的种子，也以谷物和昆虫充饥。它们在松树上筑巢，巢呈杯状，由草根、羽毛等构成。

P157小金翅鸟（1雄性、2雌性）

学名：Carduelis psaltria　英文名：Yarrell's Goldfinch

小金翅鸟的头部为黑色，虹膜为深褐色，喙为淡橙色。它的背部和腹部为黄色，背部的中央为灰色。它的翅膀和尾羽为黑色，翅膀的部分羽毛为黄色。雌鸟的头部为灰色，有不清晰的黄色过眼线。小金翅鸟性情活泼，常常成群结队地活动，每群几十只到数百只不等，繁殖期间成对生活在一起。

P158黑衣金翅鸟

学名：Carduelis barbatus　英文名：Stanley Goldfinch

黑衣金翅鸟的头部和背部为黑色，略微发黄。它的面颊为黄色，虹膜为黑色，有橙色的过眼线，它的喙为蓝黑色。它的翅膀和尾羽为黑色，羽毛的边缘为淡黄色，翅膀上有黄色的宽横纹。它的腹部为黄色。腿爪为蓝黑色。黑衣金翅鸟很少飞到地面活动，喜欢在一棵树上栖息玩耍，常活动于较高的树梢上。

P159上 赤边红眼雀（1雄性、2雌性）

学名：Pipilo erythrophthalmus　英文名：Arctic Ground Finch

赤边红眼雀的体长大约20厘米。它们经常吵吵闹闹地躲在丛林中觅食。喜欢吃草籽及植物种子，偶尔也吃昆虫。

P159下 红斑雀（1雄性、2雌性）

学名：Passerella ilica　英文名：Fox-colored Finch

红斑雀常以小群或大群出现在平地、丘陵的林缘、农耕地和庭园等的高树上活动，它们通常以植物为食，且大都是一些杂草的种子，有时也吃昆虫和谷物。

P160哈里斯雀（1雄性、2雌性幼鸟）

学名：Zonotrichia querula　英文名：Harris' Finch

哈里斯雀的头为黑色，虹膜为褐色，喙为淡褐色。它的面颊呈灰蓝色，翅膀呈褐色，有黑色的边纹及白色的横条纹，尾羽为黑色，长且上翘。它的腹部为白色，有黑色的斑点。雌鸟的头呈黑褐色，背部为褐色，有少量黑色的斑点。哈里斯雀善于鸣唱，鸣声为一串颤抖长音发自树顶。

P161褐色燕雀（雌性）

学名：Carduelis flammea　英文名：Brown Finch

褐色燕雀体长约14厘米，它的头部、背部、翅膀和尾羽均为深褐色，翅膀的边缘为淡褐色。它的虹膜为红褐色，喙为黄色。它的面颊和腹部为淡褐色，布满了深褐色的斑点，腿爪为淡橙色。褐色燕雀的巢由干苔藓、小树枝和纤草编织而成，巢通常建在柳树或桦树上。它们以柳树种子和浆果为食，夏季也吃昆虫。

P162善歌雀（1雄性、2雌性）

学名：Melospiza melodia　英文名：Song Finch

善歌雀的头部和背部为灰蓝色，头部有褐色的竖条纹和黑色的小斑点。它的虹膜为黑色，有褐色的过眼线，喙为褐色。它的翅膀和尾羽为褐色，腹部为淡褐色，有少量黑色的斑点，腿爪为褐色。善歌雀栖息于山地树林间、秋冬成大群数百只迁徙、性情活泼。活动于平原、树林、村落、园圃等处，至春季后、群体逐渐减小。图中的植物为蓝越橘。

P163莫顿燕雀（雄性）

学名：*Zonotrichia capensis*　英文名：*Morton's Finch*

莫顿燕雀的头部为灰色，有黑色宽条纹，虹膜为褐色，喙为灰色。它的背部、翅膀和尾羽为褐色，背部有少量深褐色的斑点，翅膀上有白色和淡褐色的横条纹。它的腹部为淡褐色。莫顿燕雀多栖息于山地及平原混交林中，有时也落在林间灌木丛或草地上。它们常在飞翔时鸣叫，食物以植物为主，兼食昆虫。

P164白喉燕雀（1雄性、2雌性）

学名：*Zonotrichia albicollis*　英文名：*White-throated Finch*

白喉燕雀的头部为黑白相间的竖条纹，它的虹膜为红褐色，有黑色的过眼线，喙为灰蓝色。它的面颊为灰蓝色，咽喉为白色。它的背部和翅膀为红色，背部有黑色的斑点，翅膀上有白色的条纹，尾羽为褐色。白喉燕雀把巢筑在松树或杉树上。巢十分精细，呈杯状，由杂草、细根等编织而成，外敷苔藓、蛛丝，内垫毛羽等。

P165白冠雀（1雄性、2雌性）

学名：*Zonotrichia leucophrys*　英文名：*White-crowned Finch*

白冠雀的头部为灰色，有黑色竖条纹。它的虹膜为褐色，有黑色过眼线，喙为褐色。它的翅膀和尾羽为褐色，羽毛的边缘为灰色，翅膀上的部分羽毛为黑色。它的咽喉为灰色，至尾部颜色逐渐变为淡褐色。白冠雀栖息于阔叶林，喜欢单独或成对地在高大的树顶鸣唱，声音极为优美响亮，常可由声音找到它们的踪影。图中的植物为野葡萄。

P166金冠雀

学名：*Zonotrichia atricapilla*　英文名：*Black-and-Yellow-crowned Finch*

金冠雀的头部为黑色，上面有一条橙色的竖条纹。它的虹膜为红色，面颊和腹部为淡褐色，喙为褐色。它的翅膀为黑色，有两条白色的横纹，羽毛的边缘为红褐色。它的背部和尾羽为褐色，背部有黑色的斑点。金冠雀繁殖期时雄鸟的声音非常婉转悦耳，有时甚至可持续达一分钟，非繁殖期时则不常鸣叫，也较不易发现。

P167褐色红眼雀（1雄性、2雌性）

学名：*Pipilo erythrophthalmus*　英文名：*Towhe Ground Finch*

褐色红眼雀的头部、背部和尾羽为褐色，虹膜为红色，有白色的眼圈，喙为褐色。翅膀和尾羽为褐色，翅膀的羽毛有淡褐色的边，部分尾羽的羽尖为白色。它的腹部为灰蓝色，翅膀附近由红褐色的羽毛。雌鸟的喙为灰蓝色。褐色红眼雀一般将巢建在地面，它们的卵容易受到蛇和乌鸦的袭击。现在褐色红眼雀非常稀少。图中的植物为黑梅。

P168紫雀（1雄性、2雌性）

学名：*Carpodacus purpureus*　英文名：*Crested Purple Finch*

雄性紫雀的头部、咽喉和背部为水红色。它的虹膜为深褐色，有黑褐色的过眼线，喙为黑褐色。它的翅膀和尾羽为黑褐色，羽毛的边缘为红褐色。腹部由水红色渐变至淡粉色。雌鸟的头部、背部以及尾羽为褐色，由黑色的斑点，腹部为淡灰色。紫雀喜欢群居，早晨出来活动，白天隐退到森林里，傍晚再外出活动。图中的植物为红松。

P169家雀（雄性）

学名：*Carpodacus mexicanus*　英文名：*Crimson-fronted Purple-finch*

家雀的头部和背部为褐色，它的虹膜和喙为黄褐色。它的翅膀和尾羽为黑色，羽毛的边缘为褐色，尾羽的根部为水红色。它的咽喉为水红色，至尾部变为白色，有褐色的细小斑纹。家雀食性杂，主要吃各种杂草、种子、果实、嫩芽，有时也吃昆虫、蠕虫。夏季成对营巢于茂密的树上，巢成杯状。

P170玫瑰雀（雄性）

学名：*Leucosticte arctoa*　英文名：*Grey-crowned Purple-finch*

玫瑰雀的头部为灰蓝色，有黑色的斑纹。它的虹膜为黄色，喙为灰蓝色。它的面颊、腹部和背部为褐色，背部有黑色的斑点。翅膀和尾羽为黑色，羽毛的边缘为褐色，腿爪为灰蓝色。玫瑰雀在求爱的时候，雄鸟会唱着动听的歌曲，在空中飞翔，或者响亮地拍动翅膀，以吸引雌鸟的注意。春季时玫瑰雀常发出婉转悦耳的唱声。图中的植物为紫苑。

P171交喙鸟（1雄性、2雌性）

学名：*Loxia curvirostra*　英文名：*Common Crossbill*

交喙鸟雄鸟的头部、背部和喙为橙褐色，头部和背部有黑色的斑纹。它的翅膀和尾羽为黑色，羽毛的边缘为橙褐色，腹部为黄褐色。雌鸟的羽毛有蓝灰色和绿灰色两种。交喙鸟的喙上下相互交叉，适于啄食裸子植物球果中的种子，故交喙鸟只适于生活于针叶林中。在松子严重欠收年，交喙鸟会集体大迁徙。

P172白翼交喙鸟（1雄性、2雌性）

学名：*Loxia leucoptera*　英文名：*White-winged Crossbill*

白翼交喙鸟体型似麻雀但稍大。体长约16厘米。通体朱红色，翅膀和尾羽接近黑色，部分羽毛为白色，腹部白色，面颊暗褐色。雌鸟的头部和背部为草绿色。交喙鸟一般在寒冬季节生儿育女，这是因为它们吃松子的缘故，冬天正好是球果丰收的季节。交喙鸟在"流徙"的时候任意乱飞，往往导致很多幼鸟死亡。

P173草原雀（1雄性、2雌性）

学名：*Calamospiza melanocorys*　英文名：*Prairie Lark-finch*

雄性草原雀的头部、背部和腹部为黑蓝色，背部有细小的黑色条纹。它的虹膜为褐色，喙为淡蓝色。它的翅膀和尾羽为黑蓝色，羽毛的边缘为白色，翅膀上有部分羽毛为白色。腿爪为黄色。雌鸟的羽毛为深褐色，腹部为淡褐色。草原雀成群地生活在草原地带，每群通常有60～100只。它们只在低空飞行。

P174红衣主教蜡嘴鸟（1雄性、2雌性）

学名：*Cardinalis cardinalis*　英文名：*Common Cardinal Grosbeak*

雄性红衣主教蜡嘴鸟的羽毛为红色，头部有顶饰。它的虹膜为褐色，喙为红色。喙的周围以及咽喉为黑色，尾羽为黑红色，腿爪为黄色。雌鸟的头部和腹部为橙色，背部为灰绿色，翅膀和尾羽为黑红色。红衣主教蜡嘴鸟叫声很动听，它们栖息在森林里，以昆虫和草籽为食，喜欢单独活动。图中的植物为野杏树。

P175蓝调蜡嘴鸟（1雄性、2雌性、3幼鸟）

学名：*Guiraca caerulea*　英文名：*Blue Song Grosbeak*

雄性蓝调蜡嘴鸟的头部、背部和腹部均为蓝色，虹膜为灰色，喙为灰蓝色。它的翅膀为橙色，尾羽为黑色，短小。雌鸟的头部为淡蓝色，有黑色斑点。背部和腹部为淡褐色，背部有深褐色斑点。这种鸟叫声洪亮、动听，而且生性聪明。它们喜欢吃植物种子、浆果和昆虫。

P176玫瑰胸蜡嘴鸟（1雄性、2雌性、3雄性幼鸟）

学名：*Pheucticus ludovicianus*　英文名：*Rose-breasted Song Grosbeak*

雄性玫瑰胸蜡嘴鸟的头部为黑色，虹膜为深褐色，喙为白色。它的翅膀和尾羽为黑色，有少量白色羽毛。它的背部为黑色，至尾部变为白色。它的前胸为玫瑰红色，腹部为白色。雌鸟的羽毛为褐色，有深褐色的斑点，有褐色的过眼线，喙为灰色。玫瑰胸蜡嘴鸟为热带区候鸟，栖息在森林里，以昆虫、水果和植物种子为食。图中的植物为芹叶钩吻。

P177黑头蜡嘴鸟（1雄性、2雌性）

学名：*Pheucticus melanocephalus*　英文名：*Black-headed Song Grosbeak*

雄性黑头蜡嘴鸟的头部为黑色，虹膜为深褐色，部分鸟有橙色的过眼线。它的腹部和背部为橙红色，背部有黑色的斑点。它的翅膀和尾羽为黑色，翅膀上有白色条纹。雌鸟的羽毛为褐色，有黑色的斑点，头部有黄色的竖条纹。黑头蜡嘴鸟栖息在森林的边缘，将巢建在树上或矮树丛里，巢为茶杯状。它以昆虫和水果为食。

P178傍晚蜡嘴鸟（1雄性、2雌性、3雄性幼鸟）

学名：*Coccothraustes vespertinus*　英文名：*Evening Grosbeak*

傍晚蜡嘴鸟的头部为黑色，头部的前方有一块黄色的羽毛，它的虹膜为深褐色，喙为淡黄色。它的背部为橙色，翅膀和尾羽为黑色，翅膀上有白色的羽毛。它的咽喉为深褐色，至腹部变为橙色。雌鸟额头没有黄色羽毛。傍晚蜡嘴鸟把巢建在很高的云杉上，通常它们的巢离地面6～12米。它们以虫子和水果为食，也吃植物种子。

P179夏季红雀（1雄性、2雌性、3雄性幼鸟）

学名：*Piranga rubra*　英文名：*Summer Red-bird*

雄性夏季红雀的羽毛为红色，翅膀的尖端为橙色。它的虹膜为褐色，喙为黄色，爪子为灰蓝色。雌鸟的羽毛为黄色，翅膀的部分羽毛为褐色，有橙色的边纹。雄性幼鸟的翅膀为绿色，尾羽为黄色。夏季红雀栖息在森林深处，喜欢单独活动。它们以昆虫和水果为食，最喜欢吃甲虫。图中的植物为野葡萄。

P180红唐纳雀（1雄性、2雌性）

学名：*Piranga olivacea*　英文名：*Scarlet Tanager*

雄性红唐纳雀的头部、背部和腹部均为红色，虹膜为褐色，喙为黄色。它的翅膀和尾羽为黑色。雌鸟的头部、背部和腹部为黄色，翅膀为黑色，羽毛边缘为白色，尾羽为褐色。红唐纳雀为热带地区的候鸟，以昆虫、水果和浆果为食。它们的捕食方式和捕蝇鸟相似，一般固定地站在一个树枝上等待猎物，捕食后又飞回原处。

P181路易斯安那唐纳雀（1雄性、2雌性）

学名：*Piranga ludoviciana*　英文名：*Louisiana Tanager*

雄性路易斯安那唐纳雀的头部为红色，虹膜为红色，喙为橙色。它的背部和腹部为黄色，背部中央为黑色。它的翅膀和尾羽为黑色，翅膀的部分羽毛为黄色。雌鸟的头部为黄色，背部为黑褐色，翅膀为灰色。路易斯安那唐纳雀生活在松树林里，生性活跃，叫声动听。它们以昆虫为食。

P182红翼湿地黑鸟（1雄性、2雄性幼鸟、3雌性）

学名：*Agelaius phoeniceus*　英文名：*Red-and-White-shouldered Marsh Blackbird*

红翼湿地黑鸟的头部、背部、腹部以及尾羽均为黑色。它的虹膜为深褐色，有褐色的眼圈。它的喙为灰蓝色。它的翅膀为黑色，羽毛边缘为蓝灰色，翅膀根部有一条红色和白色的宽横纹。它的腿爪为黑褐色。红翼湿地黑鸟生活在沼泽地带，喜欢群居，群居的红翼湿地黑鸟通常有500只左右。图中的植物为北美红枫。

P183红翼八哥（1雄性、2雄性幼鸟、3雌性）

学名：*Onychognathus morio*　英文名：*Red-winged Starling*

雄性红翼八哥的头部、背部、腹部以及尾羽均为黑色。它的虹膜为褐色，喙为灰色。它的翅膀根部为火红色，并有一条黄色的边纹。雌鸟的头部为黑灰色，背部和翅膀为褐色。翅膀的羽毛有白色边纹。红翼八哥栖居于平原的村落、田园和山林边缘，性喜结群，常立于水牛背上，或集结于大树上，每至暮时常呈大群翔舞空中。

P184黄鹂（雄性）

学名：Icterus icterus　英文名：Common Troupial

黄鹂又叫黄莺、黄鸟、黄栗留或金衣公子。它们体长23～25厘米。黄鹂的主要食物是有害昆虫，有时也吃果实和种子，是著名的益鸟。黄鹂的巢建在远离主干的树梢上，像摇篮一样。黄鹂是一种树栖鸟，极少在地面活动，喜欢集群，常成双成对地穿梭于树丛中。它们的鸣叫也特别动听。

P185果园金莺（1雄性、2雄性幼鸟、3雌性）

学名：Icterus spurius　英文名：Baltimore Oriole or Hang-nest

雄性果园金莺的头部和背部为黑色，虹膜为黄色，喙为淡蓝色。它的翅膀根部和背部的后半部分以及尾羽的末端为黄色。它的翅膀和尾羽根部为黑色，翅膀上有白色的条纹。雌鸟的头部、背部以及尾羽为褐色，咽喉为红褐色。果园金莺喜欢吃水果和浆果，无花果是它们的最爱。它们经常吃桑葚和草莓，令农夫们痛恨。图中的植物为火焰木。

P186北金莺（1雄性、2雄性幼鸟、3雌性）

学名：Icterus galbula　英文名：Bullock's Troopial

北金莺的头部为黑褐色，面颊为黄褐色，虹膜为红褐色，有黑褐色的过眼线，喙为淡蓝色。它的翅膀为黑褐色，有部分白色的羽毛。背部的前半部分为黑褐色，后面为黄色。它的腹部为橙色，尾羽为黑褐色。北金莺栖息在森林里，它行动活跃，喜欢群居。雌鸟非常害羞，很少鸣叫。图中的植物为黄花杓兰。

P187船尾白头翁（1雄性、2雌性）

学名：Quiscalus major　英文名：Boat-tailed Grackl

雄性船尾白头翁的面颊为蓝色，虹膜为黄色，有黑色过眼线，喙为黑色。它的头部和咽喉以及背部为黑色，翅膀和尾羽为黑色，根部为绿色。尾羽很长，呈椭圆形。雌鸟的羽毛为橙色。船尾白头翁平时三、五小群活动，繁殖季节多成对出现，到秋冬时集结群栖，普遍可见，出现于开垦地、果园、公园等环境。图中的植物为活橡树。

P188紫乌鸦（1雄性、2雌性）

学名：Quiscalus quiscula　英文名：Common or Purple Crow Blackbird

紫乌鸦的羽毛为黑色，泛着蓝绿色的光泽。它的虹膜为黄色，喙为灰褐色。紫乌鸦夜间有喜光的习性，它们喜欢栖息在高大乔木上。一般在乡间、山地的树上做碗状的窝，卵为青绿色带褐色斑点，一次生3～5个。在中国，乌鸦是不吉利的象征，人们讨厌看到它们。图中的植物为玉米。

P189褐山鸟（1雄性、2雌性、3幼鸟）

学名：Euphagus carolinus　英文名：Rusty Crow Blackbird

雌性褐山鸟全身的羽毛均为黑色,虹膜为黄色,喙和腿爪也是黑色。雄鸟的羽毛为褐色，幼鸟和雌鸟的长相相似，只有腿部和爪子颜色为灰蓝色。褐山鸟胆子很大，它们经常到牧场或者农家的院子里啄食。在冬季它们经常到潮湿的地方，例如池塘或者沼泽地带以水生昆虫和蜗牛为食。图中的植物为黑山楂。

P190草地鹨（1雄性、2雌性）

学名：Sturnella magna　英文名：Meadow Starling or Meadow Lark

草地鹨的头部为深褐色，有褐色的竖条纹。它的虹膜为黄色,有不清晰的褐色过眼线，它的面颊为白色，喙为褐色。它的背部、翅膀以及尾羽为褐色，上面有淡褐色的条纹和深褐色的斑点。它的腹部为黄色，有黑色的斑点。草地鹨喜欢成群栖息，但是到了春季鸟群会解散。鹰和蛇是它们的天敌。图中的植物为麻黄。

P191小嘴乌鸦（雄性）

学名：Corvus brachyrhynchos　英文名：Common American Crow

小嘴乌鸦属大型鸣禽，，体长约50厘米。它的虹膜褐色，体羽、嘴部及脚皆呈黑色，除腹羽外，其他部分体羽泛金属光泽的紫绿色，喙粗壮，喙基生长着密集的羽簇。喜群居生活，与其他鸦类相混生活，鸣叫时发出粗哑的嘎嘎声。一般栖息于农耕地区中，以无脊椎动物为主，但也喜欢食动物尸体，故又名食腐鸦。图中的植物为黑胡桃。

P192鱼鸦（1雄性、2雌性）

学名：Corvus ossifragus　英文名：Fish Crow

鱼鸦的体长一般为50厘米，雄鸟羽毛光滑，全身呈黑色，喙及腿爪也为黑色。雌鸟的羽毛及喙呈褐色。鱼鸦的食物以鱼和螃蟹为主，也吃谷类、浆果、昆虫、腐肉。鱼鸦喜欢群居，有时候可以成千上万只聚集在一起，然而筑巢时鱼鸦喜欢夫妻俩单独筑巢，通常用树枝筑在高高的树梢上。图中的植物为刺槐。

P193喜鹊（1雄性、2雌性）

学名：Pica pica　英文名：Common Magpie

喜鹊的头部、背部和咽喉均为黑色，它的虹膜为灰色，喙为黑色。它的翅膀为孔雀蓝色，翅膀的根部和末梢为白色。它的尾羽为草绿色，呈菱形，宽大敦厚，它的腹部为白色。喜鹊是在各地都能见到的鸟，平原、山区都有栖住，出没于山脚、林缘、村庄或城市周围的大树上、屋顶上、庄稼地上等，不见于密暗的森林内。

P194黄翼鹊（雄性）

学名：Pica nuttalli　英文名：Yellow-billed Magpie

黄翼鹊的头部为绿色，虹膜为橙色，有黄色的过眼线，喙为黄色。它的背部和咽喉均为黑色，它的翅膀为孔雀蓝色，翅膀的根部为白色。它的尾羽为草绿色，呈菱形，宽大敦厚。它的腹部为白色。黄翼鹊清早就到田地和菜地觅食；晚上栖于高大的树上。黄翼鹊是杂食性的鸟。巢多营于高大的树上，常选择近海边的木棉树。图中的植物为法国梧桐。

P195哥伦比亚鹊（雄性）

学名：Calocitta colliei　英文名：Columbia Magpie or Jay

哥伦比亚鹊的全身羽毛均为灰蓝色。它的头顶有漂亮的羽冠，虹膜和喙为灰色。它的尾羽宽大，且长，两侧的羽尖为白色。它的腹部为白色，腿爪为灰黑色。哥伦比亚鹊外形漂亮，很像中国传说中的凤凰。哥伦比亚鹊主要栖息在哥伦比亚河流域。

P196蓝冠鸦（1雄性、2、3雌性）

学名：Cyanocitta cristata　英文名：Blue Jay

雄性蓝冠鸦的头部、背部呈蓝色。它的头顶有羽冠，它的面颊为白色，虹膜为褐色，喙为灰色。它的翅膀和尾羽呈蓝色，有整齐的黑色横条纹，翅膀和尾羽的末梢是白色。它的腹部为白色。雌鸟的腹部为淡褐色。蓝冠鸦为短距离候鸟，集小群迁徙。它们主要以植物种子为食，也吃较少量的水果和昆虫。它以啄食其他鸟类的蛋著称。图中的植物为美洲凌霄花。

P197佛罗里达鹊（1雄性、2雌性）

学名：Aphelocoma coerulescens　英文名：Florida Jay

佛罗里达鹊的头顶为蓝色，面颊为白色，虹膜为褐色，有蓝色的过眼线，喙为深褐色，咽喉为白色。它的背部为淡褐色，翅膀和尾羽为蓝色，尾羽宽大。它的腹部为褐色。佛罗里达鹊一般只作短距离飞行，它们总是从一个树枝飞到另一个矮树丛。它们经常飞到地面寻找蜗牛，浆果、水果和昆虫也是它们的美食。图中的植物为柿子树。

P198灰鹊（1雄性、2雌性、3幼鸟）

学名：Perisoreus canadensis　英文名：Canada Jay

雄性灰鹊的头部为黑色，头顶为白色，虹膜为褐色，喙为黑色。它的背部、翅膀和尾羽是灰蓝色，尾羽的末端为白色。它的腹部为灰白色。雌鸟的面颊和头部为白色，头顶到背部为灰蓝色。幼鸟颜色较黑。灰鹊以腐肉为食，因此也被称为“腐肉鸟”。它们经常隐藏在丛林深处，很胆怯，容易受惊。图中的植物为栎树。

P199克拉克灰鸟（1雄性、2雌性）

学名：Clark's Nutcracker　英文名：Clarke's Nutcracker

克拉克灰鸟体长34厘米。头部、额、翼及尾羽呈黑色，尾羽末端白色。腮、喉、胸、颈及背部褐黑色，腰及尾上覆羽黑色，尾下覆羽白色，喙和脚均为黑色。克拉克灰鸟常栖息于二千米以上的针叶林，主食松子。它在飞行时尾侧可见显著的白色，鸣叫声沙哑。

P200美洲伯劳鸟（1雄性、2雌性、3幼鸟）

学名：Lanius excubitor　英文名：Great American Shrike

美洲伯劳鸟的头部和背部为灰蓝色，虹膜灰色，有黑色的过眼线。它的喙尖而且弯，雄鸟为黑色，雌鸟灰蓝色。它的翅膀和尾羽为黑色，翅膀的羽毛有白色的边纹。腹部为乳白色，有少量细小的黑条纹。美洲伯劳鸟的叫声沙哑难听，因此有人拿伯劳的叫声来比喻人说话难听。它喜欢停在空旷地的凸枝上，居高临下观察捕食昆虫。图中的植物为萎山楂。

P201傻子伯劳鸟（1雄性、2雌性）

学名：Lanius ludovicianus　英文名：Loggerhead Shrike

傻子伯劳鸟的头部和背部为灰色，虹膜为黑褐色，有黑色的过眼线，喙尖而且弯曲，翅膀和尾羽为黑色。傻子伯劳鸟以昆虫、爬虫类、小型哺乳类为主食，伯劳有将太大的猎物插在尖锐的树枝上，一块一块撕下来吃的习性，所以西方人称它为“屠夫鸟”。如果在野外见到昆虫、青蛙被插在尖锐的枝丫上，那可能就是伯劳的杰作。

P202铃雀（雄性）

学名：Vireo bellii　英文名：Bell's Vireo

铃雀的头部呈墨绿色，虹膜为淡绿色，眼周呈白色。它的喙为黑色，头部扁平，喙的根部有少量胡须状的羽毛。它的背部和尾羽呈黑黄色，翅膀灰色，有白色的边纹。它的腹部为白色，腿爪为淡蓝色。铃雀主食各种昆虫及其幼虫、卵和蛹，也吃少量植物种子。铃雀的巢很精致，巢的外围以地衣和苔藓铺成，内衬兽毛、棉花等柔软物质，很舒适。

P203黄喉雀（雄性）

学名：Vireo flavifrons　英文名：Yellow-throated Vireo or Greenlet

黄喉雀的头部和背部为绿色，虹膜为褐色，眼睛周围有黄色的眼圈。它的喙尖且稍弯，为黑灰色。它的翅膀和尾羽为黑色，有白色的边纹。它的咽喉至腹部的前半部分为黄色，后半部分为白色。黄喉雀属于捕虫林莺，叫声婉转动听。黄喉雀少动好静，喜欢单独行动。它总是小心翼翼地在丛林中搜寻食物。

P204孤独雀（1雄性、2雌性）

学名：Vireo solitarius　英文名：Solitary Vireo or Greenlet

雄性孤独雀的头部为黑蓝色，白色的眼圈与白色的过眼线相连，看起来像一个横着放的逗号。它的喙为黑色，咽喉为淡褐色，延伸至腹部。它的背部为深绿色，腹部为白色，翅膀和尾羽为黑色，翅膀泛着褐色的光泽。雌鸟的头部为黑褐色。孤独雀将巢建在地上，巢由干苔藓、鹿和浣熊的毛构成。它喜欢栖息在细长的草茎上。图中的植物为美洲罗锅底草。

P205白眼雀（雄性）

学名：Vireo griseus　英文名：White-eyed Vireo or Greenlet

白眼雀的虹膜为白色，有深褐色的眼圈和不明显的过眼线。它的面颊为黄色，头顶和喙为深褐色。它的背部、翅膀和尾羽为深褐色，翅膀和尾羽的羽毛边缘为黄色。它的咽喉为乳白色，腹部为黄色，有少量深褐色的斑点。白眼雀很少到地面活动，除非为了修补它们的巢，而从地面找纤草。它们喜欢吮吸树叶上滴下的露珠。图中的植物为苦楝树。

P206绿雀（雄性）

学名：*Vireo olivaceus*　英文名：*Bartram's Vireo or Greenlet*

绿雀的头部为墨绿色，它的虹膜呈褐色，黑色的过眼线和淡橙色的粗眉线，它的喙为灰蓝色。它的背部、翅膀和尾羽为墨绿色，翅膀的羽毛边缘为黄色。它的咽喉为白色，至腹部为黄色。绿雀很活跃，喜欢在灌木丛中来回穿越，偶尔也飞上较高的树枝。绿雀的巢都悬挂在荆棘丛中，约距地面两尺高，由粗糙的草叶构成。图中的植物为牵牛花。

P207红眼绿雀（雄性）

学名：*Vireo olivaceus*　英文名：*Red-eyed Vireo or Greenlet*

红眼绿雀的头部为蓝黑色，它的虹膜呈红色，有白色的眉线，喙为土绿色。它的面颊、背部、翅膀以及尾羽均为土绿色，翅膀和尾羽的部分羽毛为褐色。它的腹部为灰白色，爪子为蓝灰色。红眼绿雀很少飞来飞去，它们捕食主要靠在树枝上走动。只有在秋季，昆虫数量急剧减少时，它们才短距离飞行捕食。图中的植物为刺槐。

P208波希米亚朱缘蜡翅鸟（1雄性、2雌性）

学名：*Bombycilla garrulus*　英文名：*Black-throated Wax-wing or Bohemian Chatterer*

雄性波希米亚朱缘蜡翅鸟的头部有火红色的羽冠，虹膜为灰色，喙为黑色。它的背部为深褐色，至尾部变为灰色。它的翅膀为灰色，外翼有黄色羽尖，内翼为有红尖的白色。它的尾羽为黑色，末端为黄色。雌鸟没有羽冠。波希米亚朱缘蜡翅鸟胆小怕人。它们经常在树林里鸣唱，以浆果和水果为食。

P209白胸五子雀（1雄性、2、3雌性）

学名：*Sitta carolinensis*　英文名：*White-breasted Nuthatch*

白胸五子雀从头到后颈处为黑色，面颊及胸部为白色，背部及尾部为灰蓝色。分布于北美及墨西哥，常见于落叶阔叶林中。白胸五子雀喜欢在树木的枝干间跳上跳下，并会把树皮剥下，来啄食藏于树身的小昆虫，或是以树木的果实为生，它是美国庭院内常见的访客。叫声为鼻音的：“泥！泥！”。

P210红腹五子雀（1雄性、2雌性）

学名：*Sitta canadensis*　英文名：*Red-bellied Nuthatch*

红腹五子雀的头部为黑色，有白色的竖条纹。它的虹膜为深褐色，喙为黑色。它的背部为灰蓝色，翅膀和尾羽为灰蓝色，有少量羽毛为黑色，尾羽的两端呈白色。它的咽喉为乳白色，至腹部为橙色。红腹五子雀认真地检查树干的树皮和每一个洞，捕食隐藏在其中的昆虫。它们总是忙忙碌碌，吵吵闹闹地跳来跳去。

P211褐头五子雀（1雄性、2雌性）

学名：*Sitta pusilla*　英文名：*Brown-headed Nuthatch*

褐头五子雀的头部呈褐色，虹膜为褐色，有白色的眼圈，喙为黑色。它的背部及翅膀为灰色，尾羽呈黑色，尾羽的尖端为白色，腹部白色。褐头五子雀有着显著的攀爬本领。它常在树枝上用脚挟住紫杉类的种籽，用喙去啄，直到把它的仁啄出来。褐头五子雀会把卵产在铺满树皮、兽毛、羽毛的废弃的啄木鸟洞穴里。

P212加州五子雀

学名：*Sitta pygmea*　英文名：*Californian Nuthatch*

加州五子雀的头部为褐色，面颊为白色，虹膜为褐色，有白色眼圈。它的喙为黑色，长而且尖。它的背部为黑灰色，与头部的褐色泾渭分明。它的尾羽为褐色，尾羽的两侧有黄色的斑块。它的腹部为淡黄色，腿爪为褐色。加州五子雀的一生都在树上度过。它们整天不停地围着树干转，寻找树木里的昆虫。

P213黑喉蜂鸟（1、2雄性、3雌性）

学名：*Trochilus mango*　英文名：*Mango Humming bird*

雄性黑喉蜂鸟的头部为橙红色，它的虹膜为黑褐色，喙细长而且尖锐，为黑褐色。它的背部和翅膀根部为橙褐色，翅膀为灰色，尾羽为橙红色。它的腹部为灰色。雌鸟的背部为灰绿色。蜂鸟是世界上最小的鸟类，体形大小和黄蜂相似，飞行时还发出嗡嗡之声，故名蜂鸟。蜂鸟主食花蜜，也捕捉小昆虫。它活动频繁，十分活跃。图中的植物为紫葳。

P214安娜蜂鸟（1、2雄性、3雌性）

学名：*Calypte anna*　英文名：*Anna Humming bird*

安娜蜂鸟能在空中像直升飞机一般停住，所以要消耗大量的能量。而它的体形又小，这迫使它新陈代谢加强，它的体温也比其他鸟类高，达43℃，心率每分钟615次。蜂鸟的卵在鸟类中也是最小的，只有豆粒般大小，每窝只产一枚。 一只蜂鸟有1000根羽毛，是飞禽中羽毛最稠密的一种。图中的植物为朱槿。

P215红褐色蜂鸟（1、2雄性、3雌性）

学名：*Selasphorus rufus*　英文名：*Ruff-necked Humming bird*

雄性红褐色蜂鸟的头部为绿色，虹膜为黑褐色，喙为灰色，尖细且长。它的咽喉为白色，背部和尾羽为褐色，翅膀为黑色，翅膀的根部为绿色。雌鸟的背部为绿色。蜂鸟的双翼可以180°旋转，所以它既能向前飞，又能向后飞，而且能在空中停住，像直升飞机一般，因此它能自由自在地把嘴伸入花蕊中进食。图中的植物为醉蝶花。

P216象牙喙啄木鸟（1雄性、2雌性）

学名：*Campephilus principalis*　英文名：*Ivory billed Woodpecker*

雄性象牙喙啄木鸟的头顶为黑色，有红色的羽冠。它的虹膜为黄色，有红色的眼圈，它的喙为淡黄色，宽厚而尖锐。它的背部、翅膀和尾羽为黑色，颈部的白色竖条纹与翅膀尖端的白色羽毛相连接。雌鸟的羽冠为黑色。象牙喙啄木鸟是世界上最珍贵的50种鸟类之一。它们靠听觉侦测蚂蚁与木生甲虫幼虫的咬啮声。

P217北美黑啄木鸟（1雄性、2雌性、3、4雄性幼鸟）

学名：*Dryocopus pileatus*　英文名：*Pileated Woodpecker*

北美黑啄木鸟有红褐色的羽冠，虹膜为黄色，有黑色的过眼线。它的面颊为红褐色与淡黄色的条纹，喙为蓝黑色。它的咽喉为乳白色。它的背部、翅膀、尾羽以及腹部均为灰黑色，腿部粗壮有力，爪子为灰蓝色，尖锐弯曲。北美黑啄木鸟可以将树凿穿数寸深，掷出三寸长的木块。图中的植物为浣熊葡萄。

P218玛利亚的啄木鸟（1雄性、2雌性）

学名：*Picoides villosus*　英文名：*Maria's Woodpecker*

玛利亚的啄木鸟的头部为红色，虹膜为褐色，有白色的眉线，喙为灰色。它的背部为黑色，中央有灰白色的羽毛。它的翅膀和尾羽为黑色，翅膀上布满了灰白色的斑纹。它的腹部为灰白色，爪子为淡蓝色。玛利亚的啄木鸟以昆虫为生，也吃水果和果浆。它们生活在开阔的林地，农场和果园也有它们的身影。

P219绒啄木鸟（1雄性、2雌性）

学名：*Picoides pubescens*　英文名：*Downy Woodpecker*

绒啄木鸟的头部为黑色，虹膜为红褐色，有黑色的过眼线以及灰白色的眉线。它的喙为黑色，细尖且短。它的背部为黑色，中央有白色带黑点的宽竖条纹。翅膀呈黑色，有整齐的灰色横条纹。尾羽黑色，短小。雄鸟的头顶有一条红色的竖条纹。绒啄木鸟的短喙使得它们能够在小树枝上， 甚至是草茎上觅食， 对于其他的啄木鸟而言， 这的确太难了。

P220奥杜邦啄木鸟（雄性）

学名：*Picoides borealis*　英文名：*Audubon's Woodpecker*

奥杜邦啄木鸟的头部为黑色，有黄色和白色的竖条纹。它的虹膜呈褐色，喙为黑蓝色。它的背部呈黑色，有淡蓝色的花纹，翅膀和尾羽为黑色，翅膀上有白色的斑点。腹部前半部分为淡蓝色，后半部分为淡绿色，爪子呈灰蓝色。春天到来的时候，雄啄木鸟会发出响亮的叫声，那是它们在捍卫自己的地盘，警告他人不得侵犯。

P221红胸啄木鸟（1雄性、2雌性）

学名：*Sphyrapicus ruber*　英文名：*Red-breasted Woodpecker*

红胸啄木鸟的头部为玫瑰红色，虹膜为橙色，有白色的眼圈和细短的黑色过眼线。它的喙为淡蓝色。它的背部为黑色，有褐色的斑点。翅膀为黑色，有白色的边纹。它的尾羽为黑色，短而尖，有淡蓝色的斑纹。它的腹部为橙色。除了春季，红胸啄木鸟显得特别安静。红胸啄木鸟喜欢孤独，或者成双成对地旅行。

P222黄腹啄木鸟（1雄性、2雌性）

学名：*Sphyrapicus varius*　英文名：*Yellow-bellied Woodpecker*

黄腹啄木鸟的身体圆肥，它的头顶为水红色，虹膜为黑色，有白色的眼圈，黑色的过眼线延伸至背部。它的背部为灰白色，有黑色的斑纹。翅膀上有黑蓝色的羽毛。它的腹部为灰白色，尾羽为黑色，向下弯曲，羽毛尖短。黄腹啄木鸟体长约20厘米，喜欢定居在一个地区。它们在某些季节有规律地吸取树上的树汁 。图中的植物为卡罗林娜桃树。

P223黑背啄木鸟（1、2雄性、3雌性）

学名：*Picoides arcticus*　英文名：*Arctic three-toed Woodpecker*

雄性黑背啄木鸟的头顶为黄色，头部为黑色，虹膜为红色，有白色的眼圈。它的背部为黑色，略发蓝光，翅膀呈褐色，尾羽为黑色，有少量褐色的羽毛。腹部白色，有少量黑横纹。雌鸟头顶呈黑色，有白色过眼线。黑背啄木鸟的喙又尖又直又硬，舌尖有钩，舌骨有弹性，可以很灵活地伸进伸出，把蛀虫从树洞里钩出来。

P224红须啄木鸟（雄性）

学名：*hybrid-Colaptes auratus x C. cafer*　英文名：*Missouri Red-moustached Woodpecker*

红须啄木鸟生活在温带的森林中，啄木鸟跟别的鸟不一样，两个脚趾向前，两个脚趾向后，这样可以使身体保持平衡，四平八稳地攀在树上。红须啄木鸟以藏在树干中的虫子为食。它的喙尖而且坚硬，能在树干上钻孔，有趣的是，其啄虫孔能很快愈合而不影响树干的生长和发育。

P225红腹啄木鸟（1雄性）

学名：*Melanerpes carolinus*　英文名：*Red-bellied Woodpecker*

红腹啄木鸟雄鸟的头顶呈黑红色，面颊为乳白色，虹膜为红色。它的背部、翅膀和尾羽为黑色，上面遍布白色的斑纹，它的腹部呈淡褐色。雌鸟的头顶为灰蓝色。红腹啄木鸟舌尖生有短钩，舌面有粘液，所以舌能探入洞内钩捕多种树干害虫。红腹啄木鸟啄食很迅速，一次时间不到千分之一秒。

P226红头啄木鸟（1雄性、2雌性、3幼鸟）

学名：*Melanerpes erythrocephalus*　英文名：*Red-headed Woodpecker*

红头啄木鸟的头部为红色，它的虹膜为红褐色，有灰蓝色的眼圈，喙为灰蓝色。它的背部、翅膀和尾羽为黑色。它背部的后半部分到尾羽的根部为白色。腹部白色，腿爪为淡蓝色。红头啄木鸟体长约19～23厘米。活动在开阔的林地、农场和果园。它以从树木中啄出的昆虫为食，因会在死掉的树干中啄洞筑巢而出名。

P227重爪啄木鸟（1雄性、2雌性）
学名：Melanerpes lewis 英文名：Lewis' Woodpecker
重爪啄木鸟的头顶为黑灰色，虹膜为褐色，面颊为玫瑰红色，喙为灰黑色。它的背部、翅膀和尾羽为黑色，略微泛着黄色的光。雄鸟的咽喉为黑色，颈部白色，腹部和腿部呈玫瑰红。重爪啄木鸟的脚短而粗壮，二趾向前，二趾向后。尾羽刚硬具弹性，可支撑身体，这种“三足鼎立”的稳定结构，使啄木鸟能在树上自由攀登。

P228金翅啄木鸟（1雄性、2雌性）
学名：Colaptes auratus 英文名：Golden-winged Woodpecker
金翅啄木鸟的头顶呈灰蓝色，面颊呈淡褐色，虹膜为褐色，喙为灰黑色。它的头部有粉色的条纹。它的背部、翅膀和尾羽呈深褐色，背部和翅膀有灰黑色的斑点。它的翅膀和尾羽的内侧呈黄色。金翅啄木鸟善于用富有弹性的中央尾羽轴支撑身体，利用喙、舌钩取树皮下及树干中的昆虫。它冬天也兼吃一些植物性食物。

P229上 佩带翠鸟（1雄性、2雌性）
学名：Ceryle alcyon 英文名：Belted Kingfisher
佩带翠鸟是北美洲惟一一种分布较广的翠鸟，它的体长大约33厘米。上体呈蓝灰色，腹部呈白色。完全依靠食鱼为生。栖息于林间溪旁树上。

P229下 黄喙杜鹃（1雄性、2雌性）
学名：Coccyzus americanus 英文名：Yellow-billed Cuckoo
黄喙杜鹃属于攀禽，外形略似小鹰，但喙不具钩状；繁殖时，将卵产于其他鸟的鸟巢里，让其代为育雏。黄喙杜鹃喜欢开阔的有林地带及大片芦苇地。

P230红树林杜鹃（雄性）
学名：Coccyzus minor 英文名：Mangrove Cuckoo
红树林杜鹃的羽毛呈橙褐色。杜鹃鸟的产卵方式很怪异，即自己不筑巢，而将蛋产在其他鸟类的巢中，不断地以鹰的飞行方式飞翔，恐吓其他鸟类，其他鸟类误以为鹰来袭击，便匆匆离巢，杜鹃便侵占其巢，然后衔走巢中的一个蛋，它再产一枚蛋。这样其他鸟类就成了杜鹃幼鸟的养父母。红树林杜鹃以各种昆虫为食。图中的植物为苹果树。

P231马尾鹦鹉（1、2雄性、3雌性、4幼鸟）
学名：Conuropsis carolinensis 英文名：Carolina Parrot or Parrakeet
马尾鹦鹉的头部呈黄色，头顶为红色，它的虹膜为红色，有黄色的眼圈。它的喙弯曲有钩，腿较短。它的翅膀、背部、尾羽和腹部均为绿色。它的脚掌前后有双趾，走起路来样子很怪，但爬起树来却是行家。马尾鹦鹉的舌头厚而强健，能够巧妙地摆弄它们的食物——种子和水果。马尾鹦鹉的窝都设在洞中。图中的植物为苍耳。

P232带尾鸽（1雄性、2雌性）
学名：Columba fasciata 英文名：Band-tailed Dove or Pigeon
带尾鸽喜好安静，对忽然发生的事物，即使只是一个声响，它也会吓得惊慌拍翅。带尾鸽多群体性地居住或活动。在正常环境下，带尾鸽是一夫一妻制的禽鸟，配对成功后就终生不离弃。自古以来鸽子即被训练为传递信件的信差，由此可看出鸽子的记忆力之强，无论是多远的距离，鸽子总能记得回家的路。图中的植物为山茱萸。

P233德克萨斯斑鸠（雄性）
学名：Zenaida asiatica 英文名：Texan Turtle-Dove
德克萨斯斑鸠是迁徙性鸟类，身长约28厘米。它的头顶和背部呈红褐色，它的面颊呈黄色，虹膜为红色，有蓝色的眼圈。它的翅膀和尾羽为褐色，部分羽毛为白色和黑色，它的腹部呈灰色，爪子为黄色。德克萨斯斑鸠依靠在地面上寻觅食物为生，它们每天都要吃大量的小粒种子。

P234塞奈达野鸽（1雄性、2雌性）
学名：Zenaida aurita 英文名：Zenaida Dove
塞奈达野鸽的头部为橙色，虹膜和喙也呈红色。它的背部、翅膀和尾羽中央为土黄色，翅尖和尾羽两边呈灰黑色，尾羽的尖端为白色。它的腹部为橙色，爪子上有红色条纹。塞奈达野鸽通常出现于平地住家附近、农耕地、丘陵地带或河口、海边。图中的植物为番荔枝。

P235上 西威斯特鸽（1雄性、2雌性）
学名：Geotrygon chrysia 英文名：Key West Dove
野生的西威斯特鸽数量十分稀少，一般只栖息在潮湿的森林里。它的羽毛呈褐色，头顶呈黑褐色，虹膜为粉色，眼睛下有白色的条纹，喙也呈粉色。

P235下 蓝头鸽（1雄性、2雌性）
学名：Starnoenas cyanocephala 英文名：Blue-headed Ground Dove or Pigeon
蓝头鸽的头顶呈蓝色，虹膜为深褐色，有黑色的过眼线和白色的竖条纹。它的背部、翅膀和尾羽的羽毛为黑褐色，腹部呈褐色。

P236候鸽（1雄性、2雌性）
学名：Ectopistes migratorius 英文名：Passenger Pigeon
雄性候鸽的头部呈淡蓝色，虹膜为红色，有橙色的眼圈，喙为橙色。它的咽喉和腹部呈火红色，至尾部渐变为白色。它的背部、翅膀和尾羽为淡蓝色，尾羽中间长，两侧短。雌鸽的背部呈褐色。候鸽喜欢生活在院子、林地、山林等树木多的地方。它繁殖力强、性情温驯。

P237哀鸠（1雄性、2雌性）
学名：Zenaida macroura 英文名：Carolina Turtle-Dove
哀鸠的头部呈橙色，虹膜为深褐色，有灰黑色的眼圈和黑色的细过眼线，喙为暗红色。它的背部、翅膀和尾羽为灰黑色，有褐色和蓝灰色的光泽。它的腹部呈橙褐色。哀鸠走起路来昂首阔步，很神气。它们实行严格的一夫一妻制，终生不渝。即使是一方丧偶以后，也要过了很久才会再婚。

P238野火鸡（雄性）
学名：Meleagris gallopavo 英文名：Wild Turkey
火鸡又名“吐绶鸡”，最初只产于美洲。火鸡有两种，一种是平常饲养的火鸡，另一种是美洲中部的野火鸡。不论是哪一种火鸡，它们的头和脖子都是光秃秃的，没有羽毛。雄火鸡的头顶还长着红色肉瘤，喉下垂着一大串红色肉瓣；当它展开扇形尾羽，高声鸣叫时，头上的肉瘤和肉瓣还会由红色变成蓝白色。

P239上 野火鸡（雌性和幼鸡）
学名：Meleagris gallopavo 英文名：Wild Turkey
印第安人很早就知道驯养火鸡来食用。由于品种改良的结果，家禽火鸡的体形较野火鸡来得重，雄鸡可达十公斤，但是雌火鸡大约只有雄火鸡一半重量。

P239下 刀翎鹑（1雄性、2雌性）
学名：Oreortyx pictus 英文名：Plumed Partridge
刀翎鹑体形与小鸡相似，体重110～450克，头小，嘴宽短，与小鸡相比无冠，但有两根长羽顶饰，无耳叶，尾羽不上翘，尾短于翅长的一半。

P240上 冠齿鹑（幼鸟）
学名：Colinus cristatus 英文名：Welcome Partridge
冠齿鹑栖息在开阔的原野和森林的边缘地区。春天是它们的繁殖季节，冠齿鹑的食物主要是植物的种子和浆果。有时也吃树叶、根和一些昆虫。

P240下 环羽松鸡（1、2雄性、3雌性）
学名：Bonasa umbellus 英文名：Ruffed Grouse
环羽松鸡选择配偶的活动很有趣，10多只雌雄松鸡集结在一起，它们不时鸣叫。这时一只雄鸡先作各种展翅张尾的炫耀表演；然后，其他雄鸡互相追逐、争斗。观看的雌鸡会选择心仪的对象。

P241上 加拿大松鸡（1、2雄性、3雌性）
学名：Dendragapus canadensis 英文名：Canada Grouse
加拿大松鸡都生活在寒冷的地区，栖息在透光较强的林缘开阔地。它们经常在近水的向阳坡活动觅食。主要觅食果实、种子、叶、芽及昆虫等。图中的植物3为延龄草，4是腋花扭柄花。

P241下 草原松鸡（1、2雄性、3雌性）
学名：Tympanachus cupido 英文名：Pinnated Grouse
草原松鸡白天觅食时四处分散，靠互相鸣叫应答集拢。草原松鸡经过交配以后，雌鸡开始筑巢，在树根下扒一个坑，集一点枝叶和羽毛，就是它们简陋的巢。图中的植物为百合。

P242上 柳雷鸟（1雄性、2雌性、3幼鸟）
学名：Lagopus lagopus 英文名：Willow Ptarmigan
柳雷鸟羽色冬夏不同，夏羽以栗褐色为主，背肩、腰和尾上覆羽有锈褐色横斑和白色细纹，翼和腹白色。雌鸟头和颈部黑色，上体其余部分淡黄色。冬羽雪白。

P242下 岩雷鸟（1雄性冬羽、2雌性夏羽、3幼鸟秋羽）
学名：Lagopus mutus 英文名：Rock Ptarmigan
岩雷鸟一年之中伴随着春夏秋冬季节，要换四次羽毛，是鸟类中换羽次数最多者。岩雷鸟栖于山崖，食物以植物性为主，兼食各种昆虫等。该鸟善奔跑，飞行迅速，但不能远飞。

P243上 紫鹲鸪（雄性春羽）
学名：Porphyrula martinica 英文名：Purple Gallinule
紫鹲鸪的喙呈黄色，喙后端为玫瑰红色，头顶有一块蓝色的羽毛。它的虹膜呈红色，有白色的眼圈。它的羽毛呈孔雀绿色，爪子粗大，呈黄色。

P243下 鹲鸪（雄性）
学名：Thomas Bewick 英文名：Common Gallinule
鹲鸪也叫水鸪鸪。天要下雨或者刚晴的时候，常在树上咕咕地叫。鹲鸪生活在阔叶竹林和灌丛地带。杂食性动物，主要吃昆虫和草籽。

P244上 淡水秧鸡（1雄性、2幼鸟）
学名：Rallus elegans 英文名：Great Red-breasted Rail or Freshwater Marsh Hen
淡水秧鸡栖于河流附近，沼泽地带和水稻田中。它们喜欢单独或成对活动。淡水秧鸡以水栖昆虫、甲壳类、小鱼以及水藻为食。

P244下 黑脸田鸡（1雄性、2雌性、3幼鸟）
学名：Porzana carolinus 英文名：Sora Rail

黑脸田鸡的头部呈黑色，虹膜为红色，有白色或褐色的过眼线。它的翅膀、背部和尾羽呈黑色，杂有白色或褐色的斑纹。翅膀和尾羽短，只能作短距离滑翔。

P245鸣鹤（雄性）
学名：Grusamericana　英文名：Whooping Crane
鸣鹤是一种大型的珍贵涉禽。主要栖息在有水湿地或泛水沼泽，通常要求有较高的芦苇等挺水植物为隐蔽条件。鸣鹤体长可达1.2米，全身羽毛大都白色。

P246鸣鹤（幼鸟）
学名：Grusamericana　英文名：Whooping Crane
鸣鹤的雏鸟为早成鸟，出壳后即能蹒跚步行，但如无惊扰，雏鸟并不离开亲鸟远行。四、五天后，雏鸟即随亲鸟在浅水滩的草丛中觅食小鱼、蝌蚪、昆虫和各种植物嫩芽。

P247上 黑胸行鸟（1夏羽、2冬羽、3三月羽色）
学名：Pluvialis dominica　英文名：American Golden Plover
黑胸行鸟体长24厘米，多在较干的沼泽地或有短草的稻田里。冬天全身金黄色，有黑褐色斑点，故又称"金斑行鸟"，夏天则面颊、胸、腹转为黑色。

P247下 基尔第行鸟（1雄性、2雌性）
学名：Charadrius vociferus　英文名：Kildeer Plover
基尔第行鸟群栖海岸附近的休耕田，喜结队于农田或荒地中找寻昆虫、蚯蚓和植物嫩芽、根茎、种子果腹。觅食时，常以脚爪拍地，虫或蚯蚓受扰一缩动，立刻趋前啄食。

P248上 山岩行鸟（雌性）
学名：Charadrius montanus　英文名：Rocky Mountain Plover
山岩行鸟的喙为黑色，虹膜呈橙色，有黑色的过眼线。它的面颊为白色，羽毛呈褐色，翅尖有少量黑色羽毛。它的尾羽短小，腿细长，脚趾圆秃，指甲短尖。

P248下 半蹼行鸟（1雄性、2秋季幼鸟）
学名：Charadrius semipalmatus　英文名：American Ring Plover
半蹼行鸟以零星小群与其他鹬科鸟类混群活动，觅食时快速前进后啄食泥地上的小型生物。在开阔植被稀疏的卵石地筑巢，巢呈浅盘状并与环境外观极为相似，不易被发现。

P249上 笛音行鸟（1雄性、2雌性）
学名：Charadrius melodus　英文名：Piping Plover
笛音行鸟觅食时行走缓慢，常见于潮水线和积水的滩地，彼此各据一角，以昆虫、螺、软体动物为食，常与铁嘴行鸟混栖。

P249下 翻石鹬
学名：Calidris canutus　英文名：Turnstone
翻石鹬的嘴短向上翘，脚呈黄褐色，雄鸟夏羽背部为砖红色，腹部白色，雌鸟大致相似，翻石鹬喜欢以向上翘的嘴翻开石头啄食昆虫。翻石鹬多在水陆交界之处活动。

P250上 美洲蛎鹬（雄性）
学名：Haematopus palliatus　英文名：American Oyster-Catcher
美洲蛎鹬产于西半球的海岸地区。上体黑色，腹部白色，头和颈也是黑色，喙火红色，长而且根部稍细。 蛎鹬主要吃软体动物，如牡蛎、蛤、蚌类。

P250下 美洲黑蛎鹬（雌性）
学名：Haematopus bachmani　英文名：Townsend's Oyster-Catcher
蛎鹬属于涉禽，它们都有桔红色的又长又扁的喙。美洲黑蛎鹬除了腿部呈紫色，以外，全身都是黑色。 它们在海水退潮的时候吃张开着贝壳的软体动物。

P251上 丘陵矶鹞（1雄性、2雌性）
学名：Bartramia longicauda　英文名：Bartramian Sandpiper
丘陵矶鹞体长约18厘米，属于普遍冬候鸟。丘陵矶鹞停栖时，头及尾部不停地上下摆动，常贴着水面低飞而去。喜欢单独出现，见于沼泽水塘边。

P251下 红胸矶鹞（1夏羽、2冬羽）
学名：Calidris canutus　英文名：Red-breasted Sandpiper
红胸矶鹞多出现在沼泽、水田、沙洲等水浅的地带，喜欢呼朋引伴地在水中觅食。它们的警戒心很强，一有什么风吹草动，便会发出警告叫声，通知同伴们赶快离开。

P252上 紫矶鹞（1夏羽、2冬羽）
学名：Calidris maritima　英文名：Purple Sandpiper
紫矶鹞多成小群活动，常出现于水田、沼泽、泥滩及海岸等环境；它们也是性喜群居的鸟，时而可见它们和其他种类涉禽混栖，共同在湿地上觅食。

P252下 红背矶鹞（1夏羽、2冬羽）
学名：Calidris alpina　英文名：Red-backed Sandpiper
红背矶鹞身长约20厘米，走路时脚步蹒跚，整个尾部不停地上下抖动。矶鹞属于鹬科鸟类，它们的喙质颇软，富有神经，觅食时都是以长嘴在烂泥中"混泥摸螺"。

P253上 长腿矶鹞
学名：Calidris himantopus　英文名：Long-legged Sandpiper
长腿矶鹞的虹膜呈红褐色，有褐色的过眼线，和白色的宽眉线。它的喙尖长，呈黑色。它全身的羽毛为黑褐相间的斑纹，并杂有白色的羽毛。它的腿长且细，呈黑色。

P253下 小矶鹞（1雄性夏羽、2雌性）
学名：Calidris minutilla　英文名：Little Sandpiper
小矶鹞的虹膜为红色，有白色不清晰的过眼线，喙短，呈黑黄色。背部及翅膀为褐黑相间的花纹，腹部白色。小矶鹞是栖息在内陆沼泽、湖泊的少数鹬科鸟类。

P254上 红瓣足鹬（1雄性、2冬羽）
学名：Phalaropus fulicaria　英文名：Red Phalarope
红瓣足鹬夏羽色彩艳丽，头部、背部、翼、腹侧变为灰蓝色，颈部红褐色，翼面并带有红褐色泽。喙笔直尖细，脚趾间具瓣膜。冬羽期的羽毛颜色素雅，背部灰色，身体白色。

P254下 红领瓣足鹬（1雄性、2雌性、3幼鸟秋羽）
学名：Phalaropus lobatus　英文名：Hyperborean Phalarope
红领瓣足鹬体长约19厘米，小群至大群迁徙，不甚怕人。红领瓣足鹬常在河流、草泽、稻田的水面浮游觅食，动作忽左忽右，就像在跳华尔兹。

P255上 大黄足鹬（1雄性、2雌性）
学名：Tringa melanoleuca　英文名：Tell-tale Godwit or Snipe
大黄足鹬栖息于低山平原地区的河流、湖泊、沼泽、芦苇塘等开阔水域及其附近地区。常单独或成对活动。多在水边草地或沼泽地上面低空飞行。

P255下 小山鹬（1雄性、2雌性、3幼鸟秋羽）
学名：Scolopax minor　英文名：American Woodcock
小山鹬通常居住在林木茂密的丘陵，海岸、沼泽及河川等地，以水生动物为食。小山鹬保护幼鸟的方式很特别，当遇到危险时，亲鸟会用两条腿把雏鸟夹在当中一起飞走。

P256上 黑颈长脚鹬（雄性）
学名：Himantopus mexicanus　英文名：Black Necked Stilt
黑颈长脚鹬是一种美洲水鸟，生有长的粉红色腿，黑白相间的羽毛和细长的喙。黑颈长脚鹬常聚集在泥滩，把黑色的长嘴伸入水中寻找各种水生的小动物为食物。

P256下 中杓鹬（雄性）
学名：Numenius phaeopus　英文名：Hudsonian Curlew
中杓鹬体长40～46厘米。嘴黑色，细长而向下弯曲呈弧状。中杓鹬行走时步伐大而缓慢，飞行时两翅扇动得特别快，飞行有力。主要以昆虫、蟹、螺、甲壳类等动物为食。

P257上 彩鹮（雄性）
学名：Plegadis falcinellus　英文名：Glossy Ibis
彩鹮全长约60厘米。羽毛几乎都具亮古铜栗色；下背、翼和尾暗古铜绿色。繁殖季节眼先和眼周白色，远距离看是黑色。栖于热带河湖及池沼附近，有时至稻田。喜群居。

P257下 红鹮（1雄性、2幼鸟秋羽）
学名：Eudocimus ruber　英文名：Scarlet Ibis
红鹮的羽毛呈朱红色，喙长而弯曲。它们通常群聚，喜欢栖息、筑巢于宁静的松树、栎树上，涉行于清澈洁净的溪流、浅水、秧田之中，实行"一夫一妻"制。

P258树林鹳
学名：Mycteria americana　英文名：Wood Ibis
树林鹳全身洁白，头顶和颈部的羽毛为黄褐色，腿部呈黑色，喙又粗又长，有点弯曲，非常美丽。树林鹳的巢筑在高高的白杨树、栗树和松树上，山区溪流、沼泽、水田附近是它们的天堂。田螺、蟹、小鱼虾等是它们的食物。树林鹳性情温和，神态安详、高贵，受到惊吓时会哇哇大叫。

P259上 玫瑰篦鹭（雄性）
学名：Ajaia ajaja　英文名：Roseate Spoonbill
玫瑰篦鹭属大型涉禽，全长85厘米左右。玫瑰篦鹭喙长直、扁阔似琵琶，故而得名。常栖息于沼泽地、河滩、苇塘等处。涉水啄食小型动物，有时也吃水生植物。

P259下 大白苍鹭（雄性春羽）
学名：Ardea alba　英文名：Great White Heron
大白苍鹭喜欢栖息在湖泊、沼泽地和潮湿的森林里，属涉鸟类。主要食小的鱼类、哺乳动物、爬行动物、两栖动物和浅水中的甲壳类动物。它们把窝筑在树上、灌木丛或地面上。

P260黄顶夜鹭（1雄性春羽、2 10月的幼鸟）
学名：Nyctanassa violacea　英文名：Yellow-Crowned Night Heron
黄顶夜鹭性格非常隐蔽，常与小白鹭共同生活，筑巢于竹林、相思树或木麻黄林中。通常紧缩颈部，以单足伫立，维持固定不动的姿态。伫立时不会发声，通常只有在飞行中才会发出"嘎、嘎"般，粗哑且低沉的叫声。黄顶夜鹭雌雄鸟外观相同。它们以鱼、昆虫、蛙类等为主食。

P261上 小麻鸦（1雄性、2雌性、3幼鸟）
学名：Ixobrychus exilis　英文名：Least Bittern
小麻鸦属小型鹭科鸟类，全长27～36厘米。栖息于平原沼泽、湖岸和湿地，以小鱼、小虾、蛙和昆虫为食。通常在高于水面的芦苇丛、灌丛或草丛中筑巢。

P261下 绿蓑鹭（1雄性、2 9月的幼鸟）
学名：Butorides virescens　英文名：Green Heron
绿蓑鹭多单独活动，有久立等待捕捉食物的习性，飞行时两翅鼓动极为缓慢，两脚向后伸直远远拖于尾后。在浅滩觅食，晚上成群栖于高大树上。

P262大蓝鹭（雄性）
学名：Ardea herodias　英文名：Great Blue Heron
大蓝鹭是体长约1米的大型涉禽，多单独活动。它栖息在水田、河边、沼泽、海滩，以鱼、虾、蛙、蟹等水生动物和昆虫为食。捕食时，它喜欢长时间静静地站立在浅水中，待小鱼游近，快速伸颈啄捕。当大蓝鹭捕到大鱼后，会将鱼先在岸上摔死，然后吞食。繁殖期间，头后长出两条形似飘带的饰羽。集群在树上做窝。

P263上 美洲大白鹭（雄性春羽）
学名：Casmerodius albus　英文名：Great American White Egret
美洲大白鹭体长约90厘米。成鸟夏羽全身乳白色，喙黄色，肩及肩间生着成丛的长蓑羽，一直向后伸展，通常超过尾羽尖端10多厘米。成鸟冬羽背无蓑羽。

P263下 粉白鹭（1春羽、2幼鸟春羽）
学名：Egretta rufescens　英文名：Reddish Egret
粉白鹭体长60～70厘米；全身白色，眼先黄色，虹膜淡黄色，脚和爪灰蓝色；繁殖时羽背部和前颈下部有蓑状饰羽，头后有不甚明显的冠羽，喙淡黄色，喙尖黑色。

P264雪鹭（雄性）
学名：Egretta thula　英文名：Snowy Heron
雪鹭全身羽毛雪白，喙与双足为黑色，脚掌为黄色。它的虹膜为黄色，并有黄色的眼圈。生殖季节时，头部会长出漂亮的饰羽一直延伸到背部，背部的长蓑羽超过尾羽的尖端，以吸引异性的垂青。其觅食动作为双脚在水中奔跑，搜寻鱼、贝类为食。它是在沼泽、河湖、盐水池塘及河口常见的小型鹭。

P265美洲火烈鸟（雄性）
学名：Phoenicopterus ruber　英文名：American Flamingo
美洲火烈鸟又称红鹳，它们属于涉禽类。它们有"Z"字形的长颈和镰刀状的长嘴。此外它们还有一双长腿，颜色发红。它们有着独特的淡朱红色的羽毛。奇特的是无论它们的羽毛如何美丽，一经拔下就变成白色。火烈鸟筑巢一般都选择在三面环水的半岛上，有的筑于泥滩，也有的筑于水里。它们也善于长途旅行。

P266加拿大雁（1雄性、2雌性）
学名：Branta canadensis　英文名：Canada Goose
加拿大雁的喙宽短，虹膜为褐色。它的头和颈呈黑色，脸颊白色，身体和翅膀为褐色，尾羽下是白色的，雌雁的羽色稍黑。加拿大雁经常出现在路边的草地和豆科植物田中，而生长一年的草则是加拿大雁最偏好觅食的地点，次为玉米地。加拿大雁繁殖于北美，冬季南迁至诺法斯科细亚半岛。

P267上 三色苍鹭（雄性）
学名：Egretta tricolor　英文名：Louisiana Heron
三色苍鹭的背上长着漂亮的长蓑羽。它的动作轻灵而恬静，长腿单立，久久等待鱼虾出现。它在啄鱼时，善于辨别光照的方向，从不逆光行走，避免鱼被自己的投影吓跑。

P267下 白额雁（1雄性、2雌性）
学名：Anser albifrons　英文名：White-fronted Goose
白额雁体长约72厘米，是稀有冬候鸟。它全身暗褐色，有白色羽缘，胸部颜色较淡，喙粉红色，基部白色，脚橙色，腹部污白色有黑色斑块，尾下覆羽白色。

P268上 号手天鹅（成鸟）
学名：Cygnus buccinator　英文名：Trumpeter Swan
号手天鹅又称大天鹅，是现存天鹅中最大的一种，体长可达155厘米。它羽衣洁白如雪，喙乌黑，体态优雅庄重。大天鹅对爱情忠贞不渝，一旦配成对则终生为伴。

P268下 号手天鹅（幼鸟）
学名：Cygnus buccinator　英文名：Trumpeter Swan
号手天鹅栖息于多苇草的大型湖泊、池塘和沼泽地带。以水生植物为食，也吃水生昆虫和软体动物。5～6月间繁殖，巢多筑于干燥地面或浅滩的芦苇间。

P269上 小天鹅（雄性）
学名：Cygnus columbianus　英文名：American Swan
小天鹅全长140厘米，生活于江河、湖泊、水库、池塘、沼泽、高原河谷，近岸水边。在水面游泳时颈部微微弯曲，在水面拾取食物，或把头颈伸入到水中，在水下取食。

P269下 绿头鸭（1、2雄性、3、4雌性）
学名：Anas platyrhynchos　英文名：Mallard
绿头鸭的头和颈呈孔雀绿色，颈基有白色的领环；下颈、背、胸暗栗色；腰和尾上覆羽黑色，常栖于水草茂盛的湖泊、池沼，迁徙时集大群活动，以植物及动物为食。

P270北美鸳鸯（1雄性、2雌性）
学名：Aix sponsa　英文名：Wood Duck-Summer Duck
北美鸳鸯体长约47厘米，学名的意思是像新娘般美丽的水鸟。由于它们藉树洞筑巢，故又称之为美洲木鸭。北美鸳鸯是少数会飞到树上筑巢的水鸭。它们生活于森林中的沼泽、河流及池塘，以橡树子、坚果、种籽、小型无脊椎动物为食。北美鸳鸯广泛分布于北美洲、古巴等地。

P271上 针尾鸭（1雄性、2雌性）
学名：Anus acuta　英文名：Pintail Duck
针尾鸭是典型的水面觅食鸭子，常成对或结小群活动，集中于开阔水面，晚上到岸边觅食，游泳时颈直尾翘，性机警而胆怯。它通常白天在水面休息，夜间进食。飞行迅速。

P271下 蓝翼短颈野鸭（1雄性、2雌性）
学名：Anas discors　英文名：Blue-winged Teal
蓝翼短颈野鸭行走迟钝，但擅长游泳，是淡水水生性鸟类。它脚掌有蹼，颈稍短，有扁平的嘴喙。蓝翼短颈野鸭几乎失去了飞行的能力，它们只能短距离低飞。

P272上 琵嘴鸭
学名：Anas clypeata　英文名：Shoveller Duck
琵嘴鸭嘴形奇特，前端膨大呈琵琶状，觅食多在浅水中挖掘淤泥中的食物，或从水中滤食。栖息于河口、沙洲、沼泽、湖泊地带。越冬期间经常和其他野鸭在开阔水域混群活动。

P272下 赤颈鸭（1雄性、2雌性）
学名：Aythya americana　英文名：Red-headed Duck
赤颈鸭的喙与头部的轮廓属粗短型，有异于其他鸭子的流线型，头颈暗红褐色。赤颈鸭性群栖，常在较浅沼泽区、湖泊、水草丛生地带觅食、栖息，在水中潜游自如。

P273上 棕尾硬鸭（1雄性、2雌性、3幼鸟）
学名：Oxyura jamaicensis　英文名：Ruddy Duck
棕尾硬鸭常在水面倒立去啄食水底植物。喜欢开阔浅水而避免高大的水生植物。食物以嫩茎、种子为主，也吃昆虫等一些小动物。棕尾硬鸭飞得快而有力，叫声响亮。

P273下 绒鸭（1雄性、2雌性）
学名：Somateria mollissima　英文名：Eider Duck
绒鸭肥胖体重，浑身圆滚滚的。喙上有肉峰。绒鸭的雄性成鸭羽毛色彩鲜艳，引人注目。多数绒鸭都栖息在北方，在冰雪覆盖的海岸线上繁殖。

P274白斑头秋沙鸭（1雄性、2雌性）
学名：Mergellus albellus　英文名：White Merganser-Smew or White Nun
白斑头秋沙鸭是体型小约40厘米的黑白色鸭，它的喙窄而短，尖端呈钩状。繁殖期雄鸟体白但眼圈、枕纹、上背、初级飞羽及胸侧有黑色窄纹。体侧有灰色细纹。雌鸟及非繁殖期雄鸟上体灰色，具两道白色翼斑，下体白色，眼周黑色，额及顶呈栗色。 栖息于小池塘及河流等地，在树洞中繁殖。

P275上 秋沙鸭（1雄性、2雌性）
学名：Mergus merganser　英文名：Buff-breasted Merganser or Goosander
秋沙鸭别名鱼板鸭，因善于捕鱼而得名。它体长为54～68厘米，有红色或橘黄色细长且具钩的嘴，雄鸟的蹼为红色，雌鸟的蹼呈黑褐色，它们喜欢集结于湍急河流和湖泊中。

P275下 普通鸬鹚（1雄性、2雌性、3幼鸟）
学名：Phalacrocorax carbo　英文名：Common Cormorant
普通鸬鹚是冬候鸟。鸬鹚堪称是鸟类世界中的捕鱼高手，时常仗着它们高超的水性，潜到水中去追捕鱼类。鸬鹚有着发达的喉囊，可暂时贮存刚捕获的鱼。

P276双顶鸬鹚（雄性）
学名：Phalacrocorax auritus　英文名：Double-crested Cormorant
双顶鸬鹚的头顶有漂亮的顶饰，它的喙呈钩状，为黄色。它的面颊为红色，腹部的羽毛黑且泛着绿光，翅膀和背部的羽毛呈古铜色。双顶鸬鹚多出现在海边或是大型的水潭附近，常可发现它们聚集在湖岸上或是成群停栖在树梢。双顶鸬鹚停栖时，常常会将头往上仰，再挺直身体、张开翅膀来晾干身上的羽毛。

P277勃兰特鸬鹚（雄性）
学名：Phalacrocorax penicillatus　英文名：Townsend's Cormorant
勃兰特鸬鹚体长约80厘米，生活在河流、湖泊、水库、海湾等地区。快速潜泳在水中用尖端带钩的嘴捕捉鱼类。它的趾间有蹼相连，善于游泳和潜水。饱食后在陆地或树上休息时，常伸展双翅在阳光下晾晒羽毛。勃兰特鸬鹚喜结群栖息和繁殖，鸬鹚很少鸣叫，但群栖时彼此为争夺有利位置会发出低沉的"咕、咕、咕"的叫声。

P278海鸬鹚（冬季的雌性）
学名：Phalacrocorax pelagicus　英文名：Violet-green Cormorant
海鸬鹚体长约70厘米。体羽主要为黑色。头颈部带紫色光泽，嘴基内侧和眼周红褐色。繁殖期头上有两个羽状冠，胁部有白色大斑，冬羽体侧无白色斑块。它的虹膜绿色，喙黑褐色，蹼黑色。常集小群生活于海岸、河口附近，沿海小岛常见，善潜水。以鱼类、甲壳类动物为食。

P279蛇鸟（1雄性、2雌性）
学名：Anhinga anhinga　英文名：American Anhinga Snake Bird
雌性蛇鸟的虹膜呈红色，喙的上边为淡绿色，下边为黄色，咽喉红色。它的脖子细长而且弯曲，像蛇一样，所以因此得名。它的羽毛呈黑绿色，有少量白色的羽毛，蹼黄色。雌鸟的头部、脖子、前胸为橙色，腹部、翅膀和尾羽为黑色。蛇鸟栖息在湖泊、池塘和小海湾等地，它的潜水本领与鸬鹚相当，以鱼为食，也吃昆虫和爬虫。

P280白鹈鹕（雄性）
学名：Pelecanus erythrorhynchos　英文名：American White Pelican
白鹈鹕嘴巴下有弹性嘴囊，张开成袋子状。白鹈鹕通常成群围成半弧形，一起鼓动双翼拍击水面，驱鱼上浮而捕食，捕鱼时它利用大嘴连鱼带水一起吞入嘴下的大皮囊中，然后收缩把水压出。白鹈鹕的食量很大，一只成鸟平均每天可进食900～1200克，但缺食时也可以4～5天不吃东西。

281褐鲣鸟（雄性）
学名：Sula leucogaster　英文名：Booby Gannet
褐鲣鸟属于游禽，体长约70厘米，背部、翅膀和尾羽呈棕褐色，翅和尾羽的羽轴色更浓，各羽有白或棕白色羽端。前颈和胸部与上体颜色相同，腹部纯白色。喙及围眼裸皮黄绿色，喙尖长。脚黄色。褐鲣鸟是冬候鸟，群集生活于海洋，善于游泳，以鱼为食。

P282上 褐鹈鹕（幼鸟）
学名：Pelecanus occidentalis　英文名：Brown Pelican
褐鹈鹕体型比白鹈鹕小一些，体长大约107～137厘米。褐鹈鹕捉鱼的时候，从空中俯冲入水，景象颇为壮观。小褐鹈鹕靠把自己的喙伸到父母的咽喉中吃反刍食物养活自己。

P282下 黑剪嘴鸥（雄性）
学名：Rynchops niger　英文名：Black Skimmer or Shearwater
黑剪嘴鸥是个体小到中等的海鸟，它的下喙比上喙长，捕食时紧贴水面飞行，将下喙叉入水中，从而将食物送入嘴中。剪嘴鸥是热带水鸟。

P283鸥嘴燕鸥（雄性）
学名：Sterna nilotica　英文名：Gull-billed Tern or Marsh Tern
鸥嘴燕鸥体长42厘米，下体白色、上体灰白色。它的虹膜呈淡黄色，喙黑色，头顶黑色，脚黑色。以鱼为食，也吃昆虫和飞虫。鸥嘴燕鸥混在其他燕鸥族群中生长。

P284长尾贼鸥
学名：Stercorarius longicaudus　英文名：Arctic Jager
长尾贼鸥，略大于普通海鸥，羽毛多呈黑色，喜欢在山包上建窝，并且习惯于成双成对地雌雄并栖。长尾贼鸥都有自己的领地，它们一旦发现有"外族"侵入，便会誓死捍卫自己的领地。长尾贼鸥多在海岛上空飞翔，一般不到海面上活动，有时为追捕食物也会飞到离岸不远的上空与猎物周旋。长尾贼鸥的飞行能力很强，其展翼翱翔的姿势剽悍暴烈，勇猛无比。

P285普通燕鸥（雄性春羽）
学名：Sterna hirundo　英文名：Common Tern
普通燕鸥体长36厘米，喙红色，也有喙为黑色的品种，脚暗红色。夏羽时，头部在眼以上部分为黑色，眼以下为白色，身体大致为灰白色；冬羽时，额为白色，喉以下至腹部亦为白色。常和其他燕鸥混群在沙洲上休息活动，爱迎风在水域上空飞翔，时而俯冲入水捕鱼。

P286红燕鸥（雄性）
学名：Sterna dougallii　英文名：Roseate Tern
红燕鸥最大的特征，是它们鲜红色的嘴和脚，在几近白色的身上，显得格外的突出，部分个体的喙部尖端为淡淡的黑色；它们的翅膀及背部为极淡的浅灰色，头部上半部分至颈后为墨黑色；虹膜为黑色。红燕鸥的鸟巢构造非常简单，形状就像是在砾石地上的小浅坑。当温度太高时，红燕鸥会用海水弄湿自己的羽毛，为卵降温。

P287上 玄燕鸥（雄性）
学名：Anous stolidus　英文名：Noddy Tern
玄燕鸥体长约40厘米。全身大部分为黑褐色，头顶灰白色。喙黑蓝色，脚黑色。主要栖息于热带及亚热带区域之海岸或岛的礁岩峭壁上，并以海中的鱼类或软体动物为主食。

P287下 波拿巴鸥（1雄性春羽、2雌性、3幼鸟）
学名：Larus philadelphia　英文名：Bonaparte's Gull
波拿巴鸥的头部和喙呈黑色，背部、翅膀和尾羽为灰白色，翅膀的羽尖有黑色的羽毛，它的腹部为白色，脚为黄色。波拿巴鸥栖息在海边，以鱼类和软体动物为食。

p288银鸥（1成鸟春羽、2幼鸟秋羽）
学名：Larus argentatus　英文名：Herring or Silvery Gull
银鸥的背部和两翅深灰色，翼端黑色。夏羽头颈深灰色，眼周黄色，喙黄色，下喙尖端有红斑，脚粉红色，下体纯白色。冬羽头、颈部上面密布以灰褐色细纵纹，嘴黄色，栖于港湾，岛屿，岩礁和近海沿岸以及湖泊、江河附近。喜群集低飞于水面上空，杂食，以动物性食物为主。

P289大黑背鸥（雄性）
学名：Larus marinus　英文名：Great Black backed Gull
大黑背鸥除背、翼上为黑灰色，其余体色为白色。它的眼周黄色，喙黄色，下喙尖端有红斑，脚粉红色。大黑背鸥体长61厘米，是鸥类中体型最大的，它的翼展长度可达1.6米，它基本上栖息在极地附近。常杂居于灰鸥群中。常以小于10只的群体栖息，出现在河口、海岸、港湾。腐食性动物，常群集低飞水面，或停于沙滩上觅食小鱼、虾、螺蚌等。

P290上 北极鸥（1雄性、2幼鸟）
学名：Larus hyperboreus　英文名：Glaucous Gull Burgomaster
北极鸥是北极海洋上的一种普通大鸥。它浑身呈白色，虹膜为黄色，有红色的眼圈。它黄色的喙上有红色的斑点。冬天有时会迁徙到夏威夷和地中海这些南方地区。

P290下 中贼鸥（雌性）
学名：Stercorarius pomarinus　英文名：Pomerine Jager
中贼鸥生活在北极。它不但自己捕鱼还抢夺其他鸟类口中的食物。它往往是先咬住别的鸟的尾巴、翅膀，或用身体撞击，等鸟嘴中的食物一掉落，它就马上从空中接住，迅速飞走。

P291上 暴雪鹱（雄性夏羽）
学名：Fulmarus glacialis　英文名：Fulmar Petrel
暴雪鹱是海燕的近亲，它们的鼻孔像一条管子，所以被人们称做管鼻鸟。它的嗅觉很灵敏，可在3公里的距离外感觉到鱼的气味。它经常跟在船后捕食死鱼和人们抛弃的食物残渣。

P291下 灰鹱（雄性）
学名：Puffinus gravis　英文名：Wandering Shearwater
灰鹱的喙为管状，向上弯曲，喙尖向下弯。它的虹膜为褐色，有淡褐色的眼圈。它的头部、背部以及翅膀的羽毛呈黑褐色，腹部白色，脚黄色。灰鹱的嗅觉很灵敏。

P292上 凤头海鹦（1雄性、2雌性）
学名：Fratercula cirrhata　英文名：Tufted Puffin
凤头海鹦体长39厘米左右，身材短粗，脑袋大大的，头凤羽呈金黄色，背羽淡灰色，从侧面看呈扁平的红、蓝、黄三色。翅羽呈黑色，脚红褐色。凤头海鹦喜欢群栖，飞行迅速。

P292下 普通海鹦鹉（1雄性、2雌性）
学名：Fratercula arctica　英文名：Common or Arctic Puffin
普通海鹦鹉也叫大西洋海鸭，多栖息于沿海岛屿和海边，一般在离岸无人的小型岛屿之岩崖上掘洞筑巢。该鸟主要吃各种各样的海洋生物，特别喜欢食鲱鱼和沙丁鱼。

P293上 刀嘴海雀（1雄性、2雌性）
学名：Alca torda　英文名：Razor-billed Auk
刀嘴海雀陆上行走笨拙，但善于游泳及飞行，生活于悬崖上，捕食深海鱼类。卵产于海岸岩石的隙缝中。刀嘴海雀分布于北大西洋沿岸地区，为大西洋的独有品种。

P293下 凤头海雀
学名：Aethia cristatella　英文名：Curled-crested Phaleris
凤头海雀的头顶有向前卷曲的顶饰，它的喙呈红色，虹膜亦为红色，眼睛的后面有白色的细长羽。它的背部为深灰色，翅膀的羽尖有灰白色的边纹，腹部羽色较背部稍淡。

P294上 角嘴海雀
学名：Cerorhinca monocerata　英文名：Horned-billed Guillemot
角嘴海雀是善于游水和潜水的大洋鸟类。它体型中等，躯体部伸长，与颈部分界不清晰，两翼短而紧贴于躯干部，足接近尾部，通常有三趾，行走乍看颇像企鹅。

P294下 北潜鸟（1成鸟、2冬季幼鸟）
学名：Gavia immer　英文名：Great North Diver Loon
北潜鸟是高纬度地区的中型鸟类。它腿短而后移，跗跖扁平，前3趾有蹼，后趾退化；翅小而尖，多数在水面飞行，岸上行走笨拙，但在水中以敏捷的潜行技巧捕鱼为食。

P295上 红喉潜鸟（1雄性夏羽、2雄性冬羽、3雌性、4幼鸟）
学名：Gavia stellata　英文名：Red-throated Diver
红喉潜鸟颈前面有大块栗色三角斑。它善于潜水，当受惊时就迅速潜入水中，或仅将头部露出水面。飞行时，宽大的蹼尽伸在短小的尾后。红喉潜鸟以鱼类为食。

P295下 赤颈鸊鷉（1雄性春羽、2幼鸟冬羽）
学名：Podiceps grisegena　英文名：Red-necked Grebe
赤颈鸊鷉的头顶及两侧有一簇冠羽，黑色；它的前额、上胸栗红色；后颈、背灰褐色；下胸、腹白色。赤颈鸊鷉栖于多水草的水塘、江河中，以水生植物为食。

鸟类名称中西文对照及索引

A

B

C

d

F

G

H